Brain Technology in Augmented Cognition

Brain Technology in Augmented Cognition: Current and Future Trends informs engineers interested in human–computer interaction about the current state of augmented cognition. Its scope includes recent advances in electroencephalography (EEG), neural network (NN), and brain–computer interface (BCI) brain technologies.

The title explores in detail each technological approach to augmented cognition and offers conclusions to them, summarizing the work and their respective futures. Augmented cognition research often includes the use of brain technology, and this book addresses advances in augmented cognition and its applications. It details recent uses of EEG, NN, and BCI in the field and how they may augment user, researcher, and practitioner cognition. Focusing on the use of EEG for eye-tracking, NN logic, and the BCI application of motor-imagery (MI) and discussing challenges and opportunities relevant to such applications, the title is a useful introduction to the subject matter.

This is an engaging read for any student, researcher, or academic in the fields of engineering, augmented cognition, human–computer interaction, and human factors/ergonomics who will learn the basics and key concepts of augmented cognition through this simple and straightforward title.

Brain Technology in Augmented Cognition

Current and Future Trends

Suraj Sood

CRC Press
Taylor & Francis Group
Boca Raton London New York

CRC Press is an imprint of the
Taylor & Francis Group, an **informa** business

First edition published 2025

by CRC Press
2385 NW Executive Center Drive, Suite 320, Boca Raton FL 33431

and by CRC Press
4 Park Square, Milton Park, Abingdon, Oxon, OX14 4RN

CRC Press is an imprint of Taylor & Francis Group, LLC

ISBN: 978-1-032-69296-8 (hbk)
ISBN: 978-1-032-69300-2 (pbk)
ISBN: 978-1-032-69298-2 (ebk)

DOI: 10.1201/9781032692982

Typeset in Times
by Deanta Global Publishing Services, Chennai, India

I dedicate this book to my father for his inspiring, natural, and illuminating mathematics. It is also dedicated to Monte Floyd Hancock, Jr., without whose grace I would not be writing this book. I dedicate this book to everyone I have conversed with about it and its subject matter. Lastly, I dedicate this book to Human–Computer Interaction International's wonderful Augmented Cognition team and its global geniuses.

Contents

Preface

My interest in the brain began when I discovered the field of neurophysics. As a science journalist, I learned about the neurotechnological study of traumatic brain injury (TBI). This book isn't journalism, but it carries on the quest to augment our understanding and ability through use of brain technology. As concerns my being qualified to author this book, the single discipline (or, more technically, multidiscipline) uniting my educational and research backgrounds in psychology, philosophy, and augmented cognition is cognitive science.

I have had five brain scans: at least two electroencephalographs (EEGs—one during which data was accidentally not recorded!), and two fMRIs. Said brain scans were for research (the EEGs) and medical purposes (the fMRIs). My EEG scans were conducted as part of a speech decoding study, while my fMRIs were to address chronic pain (identifying tendinopathy). I have thrice contemplated buying my own EEG kit, as they are relatively inexpensive (even if messy to attach to one's head). I have also touched on EEG in my public blog, in relation to consciousness and explicit memory. Lastly, I have used a simple, single-finger biofeedback system as part of coursework in an independent study graduate course (titled by my professor and myself), "Brain and Mind Technologies".

This book was inspired by three works, in particular. It was motivated by a session I chaired at the Human-Computer Interaction International Conference 2022. This session—"Advances in Augmented Cognition – II"—included only three papers (compared with the usual six expected to fill a session at this conference). These papers were authored exclusively by students and faculty at Swarthmore College, in Pennsylvania (United States of America). I did not present in this session, but the papers' authors were collectively interested in EEG, brain–computer interface (BCI), and neural networks (NNs). I could detect the passion these researchers had in their presentations.

My journey in augmented cognition began in 2015, when I contributed to my first conference proceeding and presentation at the Human-Computer Interaction International Conference (HCIIC). With a merry band of global researchers, I conducted studies simulating the famous prisoner's dilemma, outlined a "holarchic" (subjective–objective) ontology for virtual reality (VR), and ran programs written in Python (an economics game) and BASIC (an independently written word frequency program).

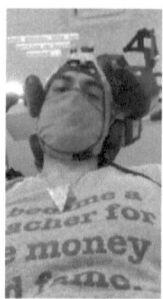

FIGURE P.1   The author took a selfie while waiting for his EEG brain scan to begin!

Since 2015, my interest in brain technology has ranged from philosophical and theoretical to commercial and clinical. My undergraduate university, the University of California at Santa Barbara (UCSB), is well-known for its cognitive neuroscience, with two titans in the field (Michael Gazzaniga and Scott Grafton, the latter of whom I interviewed as a journalist). The single discipline (or, more technically, multidiscipline) uniting my backgrounds in psychology, philosophy, and augmented cognition is cognitive science.

The aim of this book is to inform engineers interested in human–computer interaction (HCI) about the current state of augmented cognition. My experience of HCIIC is mainly with professors and technologists, with only some participants being engineers. Nonetheless, augmented cognition is relevant enough to cognitive engineering that cognitivists and engineers alike may find this book's contents interesting.

This book's scope includes recent advances in electroencephalography, neural networks, and brain–computer interface brain technologies. It also predicts and forecasts how neurotechnologies such as these will continue to be developed and applied in augmented cognition. Electrical, mechanical, and systems engineers should find all three neurotechnologies of interest. Similarly, computer engineers who work with hardware should find all three relevant, as each is discussed as a subset of human–computer interaction (NNs, though, would be of interest only to the extent that they are non-biological and instantiated in robotics). Additionally, computer engineers should be interested in the specifications needed to connect neurotechnological hardware to software and run the latter (though this book is not intended to fill this gap in knowledge). Data engineers should find the recording and transformation of EEG signals and BCI intention and the inputs fed to an NN relevant.

Sales engineers should find the ecosystem of commercial neurotechnological applications and products—such as home EEG kits—of note; they may additionally be interested in how augmented cognition (AugCog) researchers fund such projects. Mechanical engineers should be interested in how the neurotechnologies work to augment cognition through human–computer interaction. Software engineers will find in this book mentions of Python, C++, and Node.js as they pertain to EEG/BCI data processing and NN programming. Environmental engineers could be interested in resource costs associated with the development of neurotechnology and hardware waste. Civil engineers may be interested in the possibility of a neural net-powered piloting system.[1] Finally, audio engineers might be interested in the sensory substitution theorized by David Eagleman (inventor of the BCI VEST, discussed later) and should be interested in musical NN applications like the popular Shazam mobile app.[2] For nuclear engineers, it may just be worth wondering: can nuclear power help run neurotechnology?[3]

Interestingly, this book may not be as important for bioengineers. It should be for the area of neural engineering, albeit in its more technological form. Bioengineers might find brain-imaging, neural networks, and neural interfaces (viz., BCI) interesting and as extensions of their domain, but none of the three technologies focused on here directly involve changing a person's biology (or physiology). Discovering a debilitating or threatening brain condition might, however, lead to changing the brain later on. (It is debatable whether invasive BCI like Neuralink, covered briefly, changes the brain merely by being in direct contact with it.) Neural engineers should find neural networks relevant, though I have never known such an engineer as such. Chemical engineers could look to sources like Miller (2018)[4] to learn about potassium (K^+) and sodium (Na^+) (de)activations via the Hodgkin–Huxley model.[5] Neurotransmission is a chemical process, though biological NNs are only covered briefly in the present work. Finally, social engineers should find discussion of current BCI research and neuroethics of interest, as this is currently (maybe along with generative AI, which has been having its time in the sun for the last year) the most cutting-edge intersection discussed in this book.

This book could be regarded as halfway between a field manual and a reference. For those who are unfamiliar with any of the technologies discussed, consider this a primer for them in a unique context (augmented cognition). The author is aware that this book is about a specialty within a specialty: that is,

---

[1] Regrettably, structural engineers may not find much past this relevant for their discipline, so their consumption of this work would be for the civil engineer's reasons or recreation.

[2] Social and physical engineers alike should find material in this book relevant to their respective disciplines.

[3] Indeed, neurophysical processes are nuclear at the atomic level.

[4] This work also holds value for mathematicians, especially those working in probability.

[5] Probability has been interpreted as being central to computational neuroscience, per Miller. At the level of neurophysics, stochasticity is present, and quantum physics is both probabilistic and indeterminate.

it is about neurotechnologies within augmented cognition. Much knowledge about such technologies has been pulled from sources outside of augmented cognition, but when done, this is intended to supplement or feed into the more specialized presentation of them within augmented cognition. Sometimes, researchers are interested in augmenting cognition without being in augmented cognition. In this book, I include such peripheral literature to supplement the core AugCog studies using neurotech. Where I have done so, I have done so judiciously. If an article or book involves the broader project of augmenting cognition (using any of the three neurotechnologies discussed here), but is not in AugCog conference proceedings—yet, is interesting to others or in a way that AugCog could be inspired by—I have included it.[6]

It is hoped that engineers unfamiliar with the technologies discussed will feel inspired upon reading to "get their hands dirty" with them. Otherwise, useful background about them and their futures will be offered. As it happens, a book on future trends in AugCog is two-times future-oriented, given the futuristic nature of neurotechnology (as noted by the group NeuroTechX). This book will let the reader know just enough about EEG, BCI, and NNs that it will either get them started using them or (hopefully) provide them with background knowledge to enrich their existing use of such technology.

The above said, there will be a somewhat philosophical character to this book's chapters. This is partially given the author's background in philosophy, both in the university and as an augmented cognition researcher. While much augmented cognition work revolves around experiments and applications, this author's has been theoretical as well. Theory and practice are inherently entwined, so this may be deemed appropriate. This book also attempts to educate practitioners of augmented cognition on the surrounding context of the neurotechnologies they or their colleagues use and are expected to continue to (EEG, BCI, NN). This includes the history of, e.g., the convolutional neural network (CNN) and the most famous implementation of BCI outside of AugCog (yet still an augmented cognition application). Famous innovators of these technologies like Yann LeCun (convolutional NN) and David Eagleman (BCI "VEST") are covered as well, as their efforts are deemed relevant to the augmented cognition community and field at large.

---

[6] A good example of the boundary thinness between AugCog and non-AugCog works about augmented cognition is Cinel, Valeriani, & Poli (2013). This work covers EEG and BCI as neurotechnologies related to human augmented cognition. Additionally, Cinel et al. cite both a work on NN and two works explicitly using the phrase "augmented cognition" in their titles.

About the Author

Suraj Sood is an augmented cognition researcher with a decade of experience in human–computer interaction. He completed his Bachelor of Arts at the University of California, Santa Barbara, majoring in Psychology and Philosophy, and his Doctorate of Philosophy at the University of West Georgia in Psychology: Consciousness & Society. He has served as a session chair for Human–Computer Interaction International Conference's Augmented Cognition in 2019, 2021, and 2022. He has been on Augmented Cognition's Program Board since 2019. His research in HCI focuses on theory-building and simulation, drawing from psychology, philosophy, and computer science.

Acknowledgments

I thank the following online communities: the Massachusetts Institute of Technology Club of Southern California (MITCSC)—for their consistent engagement with me and enthusiasm about this book—INTJforum, Psychology Den, and Biocord. I also thank Autism Behavior Consultants for use of their technology.

Introduction

Brain technologies have the capacity to augment cognition. By this, it is meant that the affordances allowed by such technology increase our knowledge. For instance, brain-imaging via EEG—i.e., electroencephalography—tells us about the brain's activity. Still, other such technologies like fMRI (functional magnetic resonance imaging) allow us to go deeper. Of such technologies, EEG, neural networks (NN), and brain–computer interface (BCI) are of special interest in this book.

Brain-imaging neurotechnologies include EEG, as well as CT (computerized tomography), PET (positron emission tomography), fMRI, fNIRS (functional near-infrared spectroscopy), tDCS (transcranial direct current stimulation), and TMS (transcranial magnetic stimulation). fNIRS has been used in stroke recovery research; in AugCog, it has recently been used to measure participants' at-rest states to understand how varying "light illumination" (Dong, Jiang, & Liu, 2023, p. 108) affects their "cortical metabolic activity" and affect. This latter study was unique given its astronomical relevance, as illumination was measured in the context of hygiene regions of space stations and involved gravitational simulation. Though AugCog is inherently a psychological multidiscipline, given its focus on cognition, Dong et al.'s study is more psychological than the majority of neurology studies done in the field.

Though PET is not a focus of this book, it is relevant in at least one study. Okuda et al. (2003) used PET (along with cognitive tasks), finding that many of the same brain areas in the temporal and frontal lobes activated when participants thought about the past or the future. This suggests a neurological link between the cognitive processes of memory and foresight, with the latter being necessary for any predictive or forecasting work, such as that undertaken in this book. Another way to think of forecasting is that it lies somewhere between educated speculation and scientific prediction.

There are other kinds of technologies used in AugCog (the field of or a group in augmented cognition) to augment cognition. Of course, the personal computer and smart devices (phones, tablets, and watches) are used daily. Technologies include computers and the ensemble of hardware and software

DOI: 10.1201/9781032692982-1

that make up what is discussed in this book. The "smart" prefix just used refers to the fact that such devices can effectively think for their users, extending their capabilities to operate in the world. fMRI augments cognition by, for instance, alerting physicians to the state of a patient's body (including but not limited to detecting tendinopathy) and being part of high-level research university studies.[1]

The second our cognition improves as a result of interacting with one or more computers, we have participated in augmented cognition as is studied via its HCI umbrella. Using computers to understand cognition adds another layer to this, i.e., *metacognition*. Metacognition is the reason cognitive science and psychology are able to exist. If awareness is an enhancement to mindless thought or action, it can be said to be augmentative. Metacognition—awareness of thinking, or "thinking about thinking"—is by no means a topic limited to AugCog. It is even touted in public schools as a desirable phenomenon worth striving toward.

My own approach to AugCog outside of this book has not been neurotechnological. It has been decidedly theoretical, philosophical, and psychological. The flavor of this text will inevitably be colored by my background, but it will also include shades of my involvement in the broader AC community that has done interesting and successful research on neurotech. With this being said, some philosophical clarifications should be made at this point.

To augment is to enhance, improve, and make better, especially using technology. The brain is not the mind: that is, the brain does not fully explain mental activity. This book treats the mind as its own psychological domain[2], and augmented cognition as a technologization of said domain. Yet the insights drawn from use of brain technology augment our understanding of the mind, which in turn lets us optimize it. We need to understand causal cognitive mechanisms to properly augment cognition.

Another clarification is of how the concept of mind is taken up. Mind and behavior include one another. Behavior can be formalized as an aspect of mind. Mental activity can also be described behaviorally: cognition behaves in certain ways (as do affect and motivation). Behavior is also affectively and cognitively motivated (Sood, 2021, p. 21). If cognition is describable behaviorally, is behavior describable using cognitive language? Augmented cognition

---

[1] Could neurotech augment belief or logical reasoning? For instance, can it help users reach the truth more than they otherwise could?

[2] Elsewhere (see Sood, 2020), I formalized psychological mind as equaling cognition, affect, and conation. (*Conation* was Plato's term that, in contemporary psychological terms, refers to behavior, motivation, or some hybrid of both.) This formalization was also embedded in a broader formulation of *person-situation interaction*. This hyphenated construct has been studied experimentally in psychology, and I imported it into information/communication science (Sood, 2019).

happens to include behavior as a study topic, but the latter cannot be subsumed under cognition. Cognitive and behavioral psychologies are distinct intellectual traditions, though good attempts to unify them have been made (see especially Henriques, 2003).

Philosophers switch between the computational model of the mind and the "phenomenological", i.e., experiencing, mind. This book accepts both but finds the computer model a more natural fit with respect to what the fields of neurotechnology, especially, and augmented cognition, most often, aim to do. (In this book, I consider NN an example of neurotechnology. This contrasts with the comprehensive *The Neurotech Primer*, which—of the three neurotechnologies discussed in the present book—only includes EEG and BCI.) As we understand the brain and its connection to the mind, we are better able to apply knowledge of the former to augmenting the latter.

Some questions important for any cognitive science (including augmented cognition) are:

- Does the mind occupy the same spatial location as the brain?
- How distinct are mind and body?

These questions are not usually addressed in AugCog. In my research, I have taken up the philosophical "problem of other minds". The problem is whether empathic abduction (inference to the likeliest conclusion) is the closest we can get to knowing other minds exist. Successful deductive proofs of such don't seem to exist in philosophy, but AugCog is usually unconcerned with this. Most AugCog research (exceptions being much of mine, and Hancock et al., 2019) is unphilosophical.

What, exactly, is AugCog? Is it just a group of people, or is it more? If one searches for "augmented cognition" on Facebook (at least, if they were to do so at this time of writing), they might find a group profile for it (found here: https://www.facebook.com/groups/540487747185315). This profile includes

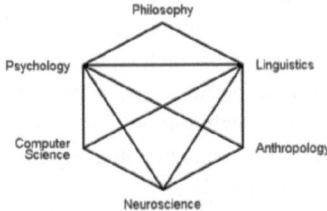

FIGURE 0.1   Diagram of cognitive scientific disciplines ("Cognitive hexagram"). The hexagram could also be expanded into a heptagram to include artificial intelligence (A.I.).

content posted by members and admins, including a job posting, a recruitment post for a past HCIIC, a post clarifying the state of HCIIC 2022, and a cover photo featuring a number of past AugCog participants taken at HCIIC 2019 (in Orlando, Florida, USA). At least in terms of social media, then, AugCog exists as a shared space to reminisce about and be discussed in relation to HCIIC and more.

Crudely (but accurately), AugCog as an area of study and application is a mishmash of computer science, psychology, mathematics, neuroscience, and more. The brand of AugCog I represent here and in general is primarily a subset of HCI, which does not fit exclusively in any one of the above cognitive science areas. In thinking about AugCog as a subset of HCI, it is most intuitive to think of technologies specifically designed for augmentation. Two such technologies are augmented reality (AR) apps and augmentative and alternative communication (AAC) devices. AR apps hit public mass in 2016 with the release of the successful mobile app *Pokémon GO*, published by Niantic. AAC devices are commonly used in special education to assist nonverbal students by allowing them to press buttons on a speech app, allowing their tablet to speak for them. Neither of these are neurotechnologies, but they illustrate that AugCog technologies in general have both entertainment value and clinical utility. In any case, cognitive science does not include all of any one discipline included in the preceding diagram.

In what sense has cognition been, or can cognition be, augmented? AugCog should include enhanced memory, foresight, learning, attention, possibly emotion (to the extent that it overlaps with cognition), consciousness, problem-solving, and decision-making. There has to be a way (or ways) in which augmented cognition is better than non-augmented, ordinary cognition. One form of augmentation is increased cognitive speed. A metaphor is sometimes used to describe the computer as a "bicycle for the mind". This is a good way to think of how cognition can be augmented: it can get us where we need to go faster, allowing us to devote ourselves, time, and efforts to other things.

There has been concern that newer technologies that automate more for the user result in a consumer who becomes less intelligent. This discourse usually revolves more around ubiquitous technologies like smartphones and not around brain technologies. Nonetheless, this points to a more general ideal of augmented cognition: that the human–computer interaction that gives rise to such cognition is smart. If technologies (including neurotechnologies) and their users can both be smart, then augmented cognition becomes a much more successful enterprise than it already is.

Brain technologies scan and image (via EEG), simulate (NN), and interface with (BCI) the brain. NetTalk is described in Varela et al. (2000) as "a grapheme-phoneme conversion machine that works by being shown a few pages of an English text in its learning phase … [that] can read aloud a new text in what many listeners consider deficient but comprehensible English" (p. 93). Tom Stafford and Matt Webb's book, *Mind Hacks: Tips & Tools for Using Your Brain*, is geared toward a popular academic audience. In it, they discuss only one of the brain technologies discussed here—i.e., electroencephalography. Yet they take it up in relation to the interesting topic of strength training, as well as to "self-caused" action.

Neurons are the basic units of the brain. They communicate via neurotransmitters, forming a collective neural network (NN). Conceived of this way, the brain is the inspiration for artificial neural networks in computer science. Often abbreviated as "neural nets" in the relevant community, these consist of two main parts: nodes (metaphorical neurons) and "edges" (segments that connect nodes). Mathematically and theoretically, the node-edge neural net draws from graph theory, a subset of mathematics. An example of nodes connecting to one another in a network is in the visual effects (VFX) software Houdini, where nodes play different roles (e.g., some are attribute nodes; others are wrangler nodes; ...). Neural nets as node-edge networks are relevant to computer science and mathematics. Neural nets proper are relevant in computational neuroscience and artificial intelligence (AI). They often employ brute-force strategies.

How many strings (in the computational, programmatic sense) occur for a given person throughout life? This kind of question requires a rigorous introspectionism, acceptance of computer science's definition of a string, and application of these both to the quantification of one's own cognition. (Of course, introspectionism requires consciousness, in this case, consciousness of one's cognition.) The reason I have focused on the quantification of cognition here is that neurotechnological measurement, in order to be relevant for AugCog, has to be informed by a well-developed cognitive science. Put another way, augmented cognition as an engineering discipline needs a mathematical basis so that the field's unit of analysis—a discrete thought, perhaps operationalized as a string to the extent that the mind is computational—is understood well enough to build an applied science on top of. We cannot improve what we do not understand (save by luck, which is not systematically replicable).

The entire field of human–computer interaction, and augmented cognition in particular, can be conceived of as a collective mind–machine interface. It is difficult to discuss a mind–machine interface without also considering the closely related subset, the brain–computer interface. Mind–machine interface is broader than BCI, the latter of which is fairly niche in its application. BCI is the most future-oriented technology to be discussed in this book, as it is shrouded in the hopes of practitioners hoping to make a difference in the lives of patients with physical disabilities. It is also very important philosophically, as the outcome of experiments involving it should shed light on the age-old question of the brain-mind connection. It is the most interesting technology of the three focused on in this book, for reasons that will be fleshed out more in its devoted chapter.

There are good reasons to augment cognition, generally. The best one, I believe, is to improve human morality. One relatively modest way to do this would be to simply fill in gaps in a person's moral compass (assuming such gaps exist). Ideally, a person could use a device like a BCI to feed them the correct moral "code" when prompted (either by said person or by detecting possible threats to morality in the user's situation: e.g., their environment)[3].

This book is geared especially toward engineers who may want to develop neurotechnologies to augment cognition. However, it is not (strictly) intended as a manual or how-to book. Treat this as a reference for background information on neurotechnologies that augment cognition, and to learn about their present state and what is likely to come next for them.

Possibility exists only in the mind. This book is an exercise in AugCog as I envision future neurotech AC trends. Such trends exist between possibility and probability: they are probable (based on the directions of current trends), and therefore also possible. This means that such future trends are likely to manifest externally, but before this, they have to be possible. The future trends I foresee in this area are first possible, then probable based on existing evidence: they are therefore more likely than not to become actual (assuming I am right). Possibility, probability, and actuality are all required for a trend to be real in a meaningful scientific—i.e., empirical—sense. If you are a neurotechnologist seeking to augment cognition, this book is most for you: you define or will define the trends discussed.

REFERENCES

Cognitive hexagram, *Wikimedia*, Accessed 23 September 2023.

Dong, J., Jiang, A., Liu, Y., Physiological and Psychological Effects of Light Illumination on Hygiene Regions of Space Stations in Short-Term Simulations of Gravity and Noise. In: Schmorrow, D. D., Fidopiastis, C. M. (eds) *Augmented Cognition. Lecture Notes in Computer Science*, vol. 14019, Springer, Cham, 2023. https://doi.org/10.1007/978-3-031-35017-7_8

---

[3] In psychological research about situations (Rauthmann, Sherman, & Funder, 2015), which are generally contrasted with persons, the physical environment could be said to be treated as a subset of situations. Specifically, environmental "cues" can be any stimuli one encounters in life, such as these words.

Hancock, M., Stiers, J., Higgins, T., Swarr, F., Shrider, M., Sood, S. A Hierarchical Characterization of Knowledge for Cognition. In: Schmorrow, D., Fidopiastis, C. (eds) *Augmented Cognition. Lecture Notes in Computer Science (LNAI)*, vol. 11580, Springer, Cham, 2019. https://doi.org/10.1007/978-3-030-22419-6_5

Henriques, G., The Tree of Knowledge system and the theoretical unification of psychology, *Review of General Psychology*, 7(2), 150–182, 2003.

Okuda, J., Fujii, T., Ohtake, H., Tsukiura, T., Tanji, K., Suzuki, K., Kawashima, R., Fukuda, H., Itoh, M., Yamadori, A., Thinking of the future and past: the roles of the frontal pole and the medial temporal lobes, *Neuroimage*, 19(4), 1369–1380, 2003.

Rauthmann, J. F., Sherman, R. A., Funder, D. C., Principles of situation research: towards a better understanding of psychological situations, *European Journal of Personality*, 29, 363–381, 2015.

Sood, S., The psychoinformatic complexity of humanness and person-situation interaction. In: Arai, K., Bhatia, R. (eds) *Advances in Information and Communication. Lecture Notes in Networks and Systems*, vol. 69, Springer, Cham, 2019. [Also presented at the Future of Information and Communication Conference 2019] https://doi.org/10.1007/978-3-030-12388-8_35

Sood, S., The Platonic-Freudian Model of Mind: Defining "Self" and "Other" as Psychoinformatic Primitives. In: Schmorrow, D. D., Fidopiastis, C. M. (eds) *Augmented Cognition. Theoretical and Technological Approaches. Lecture Notes in Computer Science*, vol. 12196, Springer, Cham, 2020. [Also presented at the 22nd International Conference on Human-Computer Interaction, pp. 76–93]

Sood, S., Holarchic HCI and Augmented Psychology ("AugPsy"). In: Schmorrow, D. D., Fidopiastis, C. M. (eds) *Augmented Cognition. Lecture Notes in Computer Science (LNAI)*, vol. 12776. Springer, Cham, 2021. https://doi.org/10.1007/978-3-030-78114-9_22

Varela, F., Thompson, E., and Rosch, E., *The Embodied Mind*, The MIT Press, Cambridge, 2000.

How Can Brain Technology Augment Cognition?

1

1.1 INTRODUCTION

The basic question of augmented cognition is: Does technology boost thought? If so, how? Traditionally, "technology" in this formulation can be replaced with "computer". If using a computer can improve thinking, it is a good candidate for inclusion in augmented cognition. Augmented cognition—colloquially abbreviated as "AugCog"—"is...focused on accelerating the production of novel concepts in human-system integration and includes the study of methods for addressing cognitive bottlenecks (e.g., limitations in attention, memory, learning, comprehension, visualization abilities, and decision making) via technologies that assess the user's cognitive status in real time" (Schmorrow, 2014, July 4). A primary aim of the field "is to research and develop technologies capable of extending, by an order of magnitude or more, the information management capacity of individuals working with 21st Century computing technologies".

AugCog science, technology, research, and development also observe a user's state, recording neuropsychological and physiological data in real-time to enhance said user's carrying out of a task. The thrust of AugCog is largely leveraging technology and science to engineer human psychology (cognition).[1] AugCog also consists of "a set of theories, principles, and computational sys-

---

[1] Sometimes, affect and conation (motivated behavior) are also involved.

 DOI: 10.1201/9781032692982-2

tems to support and extend human cognitive abilities in real time by taking into explicit consideration well-characterized limitations in people's attention, memory, problem solving, and decision making" (Stanney, Winslow, Hale, & Schmorrow, 2015, p. 329).[2] This definition was (presumably) given for an audience of those working in psychology.

However, before we can augment cognition scientifically, we must be aware of AugCog's place in our daily lives. An act as simple as reading intellectual discourse in a chat room can augment cognition (as a function of human–computer interaction), but we aren't normally aware of just how much it does so. Bringing the above discussion down to earth, I was reading a discussion of neural dynamics immediately prior to writing the above paragraph. I next performed a Google search and found a good definition of AugCog (albeit not provided by an explicitly AugCog resource).[3] I was then inspired and mentally stimulated enough to begin typing the last and current paragraphs.

The above is surely AugCog in action: albeit, absent of any neurotechnological measurement, stimulation via deep brain stimulation or BCI, or any neural network in sight. But if the brain can be discussed in such a manner online, then the Web and perhaps its parent technologies—the Internet and computer—could be considered neurotechnological in some looser sense than normally used in this book. Certainly, the Internet and its subset, the World Wide Web, may be responsible for the observed Flynn effect, an increase in intelligence quotient (IQ) of later human generations. Access to and participation in the World Wide Web may be responsible for this, and if so, the Web is a direct application of augmented cognition.

This book is about cognitive augmentation, a kind of engineering with the goal of optimizing and enhancing thought. The engineering in question here is both old and new. Inner engineering is a spiritual practice drawing from Hindu yoga; cognitive engineering cannot be newer than cognitive psychology and cognitive science. Augmentation is present in popular culture. The term even has its own Fandom webpage as part of the Deus Ex game Wiki (Deus Ex Wiki). It is interesting to note the differences between AugCog and the definitions of augmentation on this website, where cognitive augmentation is not even discussed. This site instead defines augmentation in terms of its pharmaceutical, nanotechnological, and cybernetic variants. Any of these could be relevant to AugCog, especially the latter two. Nanotechnological neural chips are conceivable, and cybernetic BCI is seen in media—e.g., prosthetic arms that respond to their user's intent through movement.

---

[2] This work also includes relevant background on EEG and BCI (as well as references on NNs).

[3] By the way, here is the first resource whose link summary I read before writing this and the previous paragraphs: https://www.smartbrainpuzzles.com/blog/mental-stimulation-for-mental-health/#:~:text=So%20What%20Exactly%20is%20Mental,or%20even%20socializing%20with%20people.

Augmentation in the context of this book is only a function of human–computer interaction (and, for others outside of my area of expertise, human factors and ergonomics). There is no clear mathematics of cognitive augmentation. I have spent a great deal of thought, time, and effort developing a mathematics for HCI (VR) and psychology. Cognitive modeling can be mathematically sophisticated; my approach to cognitive math is simpler. Mathematics accompanying cognitive engineering and science would be appreciated. But first, we must be clear on what cognition is.

Essentially, cognition is the technical term for thought. But thought comes in many varieties: for some in cognitive psychology, affect (roughly synonymous with, if not closely related to, emotion and feeling) is a kind of cognition. I believe this oversimplifies matters and possibly even commits a category error. Nonetheless, an important consideration is how AugCog treats emotions and their role in, or influence on, cognition. Are emotions necessary for thinking? Should they inform some cognitive processes—like decision-making—but stay out of others? Aside from the vagus nerve,[4] do neurotechnologies have any relevance to emotions (including their processing) or the heart? And can we distinguish affect from instinct in any meaningful way, in the neurotechnological AugCog context?

Cognition (which I formalize as C) also includes imagination, memory (formalized below as Mo), foresight (Fs), and (also the purview of economics) decision-making (D-M). So, cognition could at least be mathematized as a function consisting of these faculties.

$$C = f(\text{imagination,}[5] \, Mo, \, Fs, \, D\text{-}M) \tag{1.1.1}[6]$$

Cognition is often colored by affect (A), and can even be converted into behavior:

$$C_A = B \tag{1.1.2}$$

Cognition is also of varying degrees of consciousness. This is closer to the math I have developed, drawing from Platonic philosophy and Freudian theory.

$$M = f(A, \, C, Mv)_{(U\text{-},sb\text{-})Cs} \tag{1.1.3}$$

---

[4] The polyvagal theory forwarded by Dr. Steve Porges gained popularity due to its link between the heart and the nervous system. The vagus nerve is an interesting topic since most discussions of the nervous system focus on the central nervous system (CNS), including the brain, and the peripheral nervous system; the heart is usually left out.

[5] Is what the imagination produces necessarily possible in actuality?

[6] Here, I follow the notation used in previous publications of an attempted mathematical psychology.

Equation (1.2.3) reads: Mind M is a function of unconscious (UCs), sub-conscious ($SbCs$), and conscious (Cs) affect A, C, and motivation Mv. Such math makes cognitive augmentation a much more straightforward enterprise. All that needs to be done is to "plug in" this math into whatever cognitive augmentation device is of interest.

Augmented cognition can be defined in four ways. As described above, it is a multidisciplinary field interested in enhancing and extending thought via computer technologies. AugCog usually refers to a group of researchers including academics and industry professionals who present and have their work published, especially in conference proceedings. My experience in augmented cognition has been as a research assistant, independent and group researcher, assistant project manager, co-director, and program board member for a special interest group. I have been part of a team within Augmented Cognition, which itself operates partially (i.e., not exclusively) under the Human–Computer Interaction International Conference.[7] AugCog has a social media presence and meets when there are work, directional, or action items to be clarified. Augmented cognition is also a phenomenon of interest in these contexts.[8]

What, in addition to what has been stated above, is cognition? Is it neurophysiological? Does it include affect (or emotion)? My favored view is that it is simply thought that we operationalize. Then is it empirical? Thought is not so, ordinarily. Via cognitive neuroscience, we suppose that brain patterns correlate with discrete thought processes—but the nature of the brain and thought is different at the very least in the words we use to describe each.[9] Thought is also paired to some extent with technology.

Neurocognitive research has come a good way. One need look no further than improved ethics in what should or should not be done to the brains of patients to confirm this. Since last century (when brain lesions were conducted), we have a better idea of how to handle brains with care. The treatment of brain disorders is the most clinical form of augmenting cognition. Any such effort should raise cognition at least to a baseline level of functioning. Past this, augmented cognition takes on a more positive quality, raising it from baseline up to extraordinary functioning. Such AugCog should be used

---

[7] An augmented cognition community also exists on LinkedIn at the time of writing, named the Augmented Cognition Technical Group (ACTG) of the Human Factors and Ergonomics Society (HFES). HFES has been called the sister of HCI.

[8] Terms like *neurodiversity* and even "neurotalent" denote cultural movements without any necessary appeal to the brain, proper. Given this, could such movements—which often manifest online—be edge cases of HCI-augmented cognition?

[9] The difference in my view is greater than just language. In Sood et al. (2019), I distinguished between biological and psychological realities. The brain is a physical, chemical, and biological entity (and for some, computational). The mind is psychological.

to solve difficult problems and create unique visions (of improved states of affairs) to then implement using extended performance.

Cognition consists of many faculties. It includes phenomena like memory, perception, learning, and attention. In cognitive psychology, attention is discussed in terms of a metaphorical *attentional spotlight*, a technical term referring to our capacity (or tendency) for selective attention. Each of these subsets of cognition can be divided further. Learning may be implicit (automatic) or explicit (conscious). Implicit learning includes habits that we acquire through repetition. A topic of interest in augmented cognition is *metacognition*, or thinking about thinking. Metacognition conceived of in this way received its own session semi-recently at the Human–Computer Interaction International Conference. This conference takes place annually and includes a prominent global context—one of, if not the, largest—for sharing augmented cognition findings.

Memory is usually broken down into long-term and short-term forms. Memory also implicates emotion (which represents its own topic, discussed within yet also autonomous from cognitive science). Perception is usually thought to be sensation-plus: it engages the five commonly conceived senses of taste, sight, hearing, smell, and touch. Perception adds an interpretive layer to this sensory ensemble: thus, it is a conscious process, and is more directly relevant to cognition than is sensation.

Cognition is both structural and processual. Its processes can include memory, attention, and so on; further examples are problem-solving, language, motor activity, and foresight. Language as cognition involves encoding, acquisition, and lifespan development. Motor activity is perhaps an unusual inclusion, but as will be seen later in this book, it is taken up in augmented cognition research.

Augmentation of learning, memory, perception, and emotion are each an example of augmented cognition. AugCog is often also interested in augmenting cognition as evidenced by extending a person's performance. Theoretically, this begs for an expansion of the view of cognition that it is bounded within one's skull, identical with the brain. Andy Clark argued for the extended mind thesis, which seems a good way to situate this aspect of AugCog philosophically speaking.[10] Adding computer technology to this mix, any such tech that enhances our cognitive faculties is fair game for the field. Augmented cognition has been described as, and indeed is, multidisciplinary, drawing research from the brain, cognitive, and computer sciences. It is an opportune time to

---

[10] What cannot be done from a philosophical or psychological viewpoint is subsuming behavior under cognition. (This would violate Plato's view of the mind.) It is true that we can think about action, but action also influences thought, suggesting a reciprocal relation. More precisely, cognition is extended via AugCog to be linked with behavior. What results is then a hybrid of cognition and behavior, and perhaps next the transformation of cognition into behavior.

pause on the relation between augmented cognition and the brain, given the decade of the brain has just passed: I measure this in terms of the start of the Brain Research Through Advancing Innovative Neurotechnologies® (BRAIN) Initiative.

The BRAIN® Initiative sought to do for the brain what the Genome Project did for our genetics, but by mapping our brain cells (rather than the DNA inside of cells). Whether one considers 1990 or 2013 the start of the decade of the brain—the latter year is when the BRAIN® Initiative began—the turn of the millennium has seen some interesting ties formed between augmented cognition and neuroscience. These ties are the subject of this book, with a particular focus on the technological side of things.

The reader will note that I have included numerous sources in this book outside of the field of AugCog. This is simply because neurotechnological use for the purpose of cognitive augmentation of some kind is widespread enough to allow this. It may also lead to fruitful cross-pollination in the future between the field of AugCog and studies implicitly involving augmented cognition as a phenomenon.

1.2 BRAIN TECHNOLOGIES AND THE MIND

The relation between brain technologies and the mind is simple enough. Using our minds, we conceive of technologies that will hopefully allow us to discover important things about who we are. Neurotechnologies help us understand the brain and complement cognitive research by giving us a picture of our neuropsychology. Undoubtedly, neurotechnology is an empirical discipline. Augmented cognition is primarily technological (and experimentally empirical), but it has its theoretical arm. Theoretically, we augment unconscious-conscious cognition.[11]

How can we measure unconscious thought, let alone augment it? One answer offered by Freud is that the unconscious reveals itself via the subconscious. Dreams in particular were considered a gateway through which unconscious content (e.g., thought) passes into consciousness. Freudian theory is not usually taken up in AugCog (either implicitly or explicitly), with the exception of my own theoretical work. But dreams are in principle relevant to AugCog given their necessarily cognitive quality. Dreams are the sort of thing

---

[11] Sensation and perception are also facets of the mind. Could these be sensibly augmented?

that could be influenced via lucid dreaming and waking-world stimuli (such as music). So, perhaps they could be augmented.

In past work, I contributed to augmented cognition what I termed the Platonic-Freudian[12] model of mind. Cognition can be unconscious, subconscious, conscious, or even superconscious. Further, humans may have cognitive states, traits, or processes. This gives birth to a matrix of cognitive ontology. Unconscious, subconscious, conscious, and superconscious cognitive states, traits, and processes can be said to exist. Mathematically, this yields 12 cognitive possibilities:

- Unconscious cognitive states
- Unconscious cognitive traits
- Unconscious cognitive processes (e.g., intuition)
- Subconscious cognitive states (e.g., rapid eye movement (REM)-dreaming)
- Subconscious cognitive traits
- Subconscious cognitive processes
- Conscious cognitive states
- Conscious cognitive traits
- Conscious cognitive processes (e.g., deliberative problem-solving)
- Superconscious cognitive states
- Superconscious cognitive traits
- Superconscious cognitive processes[13]

This makes the terrain for augmented cognition potentially vast. Can brain technologies augment all of these kinds of cognition? A goal of augmented cognition could be to either raise subconsciousness or lower superconsciousness to regular consciousness. Consciousness is a specific aspect of the mind. Scientists from neuroscience and computer science gathered and wrote out their stance that consciousness is something shared by all organisms with working central nervous systems (CNSs) (Low, Panksepp, Reiss, Edelman, van Swinderen, & Koch, 2012). Low et al. stated:

---

[12] Sigmund Freud was a neurologist before his interest turned to the mind, in its unconscious-to-conscious forms. Another interesting perspective for this book is the view of him as the prototypical AI theorist (Minsky, 2013).

[13] Given their prefixes, *subconscious* and *superconscious* may logically be considered opposites (where the former refers to partial consciousness, and the latter refers to hyper-consciousness). This potentially opens the door to a fifth state of consciousness to be contrasted with unconsciousness. It may be that regular consciousness fills this gap, but it is worth keeping the door open for possible future developments in our understanding of consciousness and all of its variants.

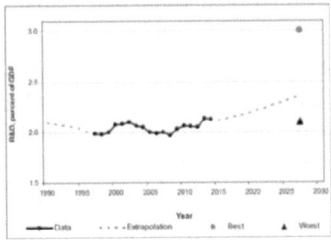

FIGURE 1.1 Global science and technology R&D: gross domestic product (GDP) across 40 years with forecasts made for the current decade (Glenn et al., 2019, p. 46).

non-human animals have the neuroanatomical, neurochemical, and neurophysiological substrates of conscious states along with the capacity to exhibit intentional behaviors… Non-human animals, including all mammals and birds, and many other creatures, including octopuses, also possess these neurological substrates.

Thus, for Low and his colleagues, any living mammal, bird, or octopus possesses the physical, chemical, and biological constitutions that are believed to correspond with behaviors agreed upon by its observers as correlating positively with volitional action. It may be that someone is either not aware enough ("cognizant" might be a better word choice, in this context) or is too aware in some task. Conscious volition is an important part of BCI research and application in a way that will be described in Chapter 4.

This book assumes no foreknowledge of what constitutes a trend. Engineers reading may (have) work(ed) with or know about the neurotechnologies discussed. Background knowledge concerning such technologies will be provided for review, or in case readers are unfamiliar with how they work.

1.3 THREE CASE EXAMPLES

The three neurotechnologies I will be primarily engaging with in this book are electroencephalography (EEG), brain–computer interface (BCI), and neural networks (NN). For the more careful reader, ethics is discussed well with respect to neurotechnology in Müller and Rotter (2017). These authors engage with philosophical notions of spirit and soul, making their ethical discourse far deeper than much of such discourse in philosophy (ethics' most basic surrounding discipline). Opinions vary with respect to the adoption of technology.

Augmented cognition technologists should be aware of such considerations as we continue contributing to and creating the technologies of tomorrow.

Most neurotechnology in augmented cognition is used "indirectly". This means that the use of such technology yields insights that is then applied to augment understanding and user performance. The three technologies covered mainly in this book do not operate at the physical level of the human brain, instead covering parts of the body or existing purely as software. This does not mean more directly applied neurotechnologies are irrelevant to augmented cognition: deep brain stimulation and neural chips are examples respectively of clinical and more exploratory AugCog applications. Brain stimulation features in a philosophical thought experiment regarding the nature of the brain-mind relationship. It is argued that if stimulating a certain brain region affects the mind (e.g., perceptual cognition), then the mind is a physical entity closely related—if not functionally equivalent or identical—to the brain. This is the most convincing argument for the concept of the mind merely being a different way of describing the brain (or nervous system). Irrespective of this, a cautionary ethic should be adopted before applying such "intrusive" technologies to humans.[14]

So: how can these three neurotechnologies augment cognition? EEG augments cognition by rendering clear brain region activity of individuals engaged in specific tasks. More specifically, event-related potentials (ERPs) measured at various electrodes may be compared during recall memory research, potentially allowing for neurocognitive localization. BCI augments cognition by interpreting the wearer's conscious volition and turning it into outward behavior. Neural networks augment cognition by acting as a brain, enabling artificial intelligence—and, thus, possibly creating cognition—in either a computer system or organism. Simulations performed by neural nets also augment cognition by telling us what could result from an artificial brain operating in unique, environmentally embedded, and goal-directed ways.

Detecting trends takes a patient expert eye. The trends discussed in this book are primarily technological. However, these could be segmented into sub-trends, including aesthetic and functional kinds. Functional trends of specific technologies should stay largely stable, but they are usually updated incrementally to improve functionality. This especially includes ergonomic considerations, since technologies often trend toward becoming easier to use for non-specialists. Aesthetic decisions may be made to make technologies look more appealing (e.g., cooler). Another consideration pertinent to the evolution of current trends into future ones is how long and strong a trend of the former variety needs to be to become a future trend. There may be some necessary and sufficient conditions for a trend we observe now to extend into the future.

---

[14] Some neurotechnologies aim to alleviate symptoms, as deep brain stimulation does for major depressive disorder.

My own AugCog research has had themes that I identified prior to starting this book. These are as follows:

- 2015—emotion
- 2017—agency
- 2018—games
- 2019—reality
- 2020—mind
- 2021—language (and happiness)
- 2022—love
- 2023—religion and spirituality

From 2020-2021, I also conducted thematic analysis on case studies for my doctoral dissertation. The themes of interest above were coded for in the biographical material of people of interest (environmentalists). However, the first trend that I noticed (though I called it a "wave") was field theory, featured in Hancock et al.'s 2015 and (one of their) 2020 papers. Does this give a clue for this book's context—forecasting neurotechnological trends in AugCog?

Trends may deviate from what is expected. They are patterns observed throughout time. Technological trends are humanmade. This means such trends are a function of what makes AugCog neurotechnologists apply and create the relevant technologies that they do. Also important to consider are relations between trends, such as those that reinforce one another and those that compete. Funding of neural research activities in AugCog is expected to reinforce increased use and/or development of neurotechnologies in this field. In contrast, funding of scientific research not pertaining to neurotechnology or AugCog could be considered a competitive trend, as it means less room for AugCog neurotechnologists to invest in the kind of work they do.

Can trends be predicted in addition to being forecasted? In social science, prior to the wave of big data, it was advised that a sample of $N =>30$ must be included for an empirical study's results to possibly be statistically significant. Using this as a baseline, if we had 30 years of AugCog research to study, one could perhaps reach a comparable level of significance in a prediction. But prediction is closer to hypothesizing than to a study's findings. Still, hypotheses should be based on the existing state of data at a given time, so this method of prediction could still work. If a prediction is more solid than a forecast, then if achievable, the former should be preferred for predicting the state of an empirical discipline like AugCog.[15]

---

[15] There are a few problems with this approach. First, we are firmly in the era of big data. Second, big data is close to HCI given their mutual interests in digital technology (with AugCog in particular leveraging such technology for cognitive advancement). However, AugCog is not

My experience in AugCog began in 2015 (nine years ago). This means I have not had direct contact with 30 years of conference proceedings to predict its future trajectory in the manner described above. Still, general trends in the smaller sample I have available could be enough to give a good idea of AugCog neurotechnology usage likely to appear in the short run.

Themes are related to trends. Trends can be thought of as themes cropping up for periods of time, disappearing, and possibly reappearing. Trends are defined by more stakeholders investing in them than in things not considered "trendy". There is a subjective aspect to trends: to understand this, one can look to the close relative of them, fads. Fads are essentially informal trends existing in culture. Both fads and trends come and go, though the former is more fleeting while the latter is more serious or objective (a fad could be something as superficial as many people consuming an entertainment brand's products, e.g., Pokémon trading cards). Fads and trends have some economic factor(s) supporting their spread, such as a supply–demand equilibrium—this is accompanied by heightened group interest. A topic or research interest is trending if it attracts a substantially larger number of people than topics or interests that are less popular.

It is easy to see why neurotechnology is a broader trend, given this century's thematic focus on the brain. For AugCog, it is due to the close relation between thought and the brain. It is hypothesized widely that investing in understanding and improving neurotechnology will lead to augmented cognition. Enough scientists, theorists, and industrialists in AugCog take this hypothesis seriously that neurotechnology could properly be considered a trend within the thematic area of AugCog.

A forecast is broader than a scientific prediction. Predictions must be warranted by the evidence gathered to date. Such prediction is a primary function of science, though it is usually confined to specific hypothesis-testing in studies. In Agile software teams, with "sprints" lasting two weeks, it is easy to predict when a project will be complete. This may be helpful for forecasting neural net and EEG software development. An example of a well-known trend is the "hockey stick" graph that shows a rise in global temperature, especially between 1950 and now.

Correctly predicting or accurately forecasting future trends is perhaps the closest we can get to knowing the future. More technically, we *abduce* the future using these methods: i.e., we infer it probabilistically. We only know it in a purely empirical sense once it has come to pass, at which point it becomes

---

old enough that we have a very large number of proceedings to use as data points: a maximum sample size of 75 could be speculatively set for the ideal case. Third, the recommended minimum sample size for a study with human subjects may be different for the kind of study done in this book, which uses non-human data (viz., conference proceedings).

our present. In future studies, forecasting is sometimes done to envision multiple possible futures (as well as an optimal one). I have had a foot in a network of futurists—The Millennium Project—since 2016. Though I do not employ their methods in this book, one worth mentioning is the real-time Delphi (RTD) used to collect and aggregate "expert opinions" (The Millennium Project, 2023). A benefit of this is that expert consultants can project the future state of an area (like neurotechnology) dynamically, such that trend forecasting becomes a continuous, collaborative process that incorporates new data.

Forecasts can be made for the short term (such as in weekly weather forecasting) or over a longer period (e.g., quarterly financial forecasts). Trends might exist in the past (e.g., the direction of global climate in the last century) or the future (e.g., next week's weather). The surest way to be able to forecast about technology in the short run is to develop technology oneself or know others who do, to know development schedules, and to know how competent, driven, and reliable such tech developers are. Within AugCog, spending time in the community as I have is a good help for such forecasting. Otherwise, following projects on GitHub or knowing update needs and schedules helps with general software forecasting.

We can forecast trends both naively and knowledgeably. Naive forecasts can be treated as hypotheses to be tested using a future state of increased knowledge about the space in question (here, AugCog neurotechnology). Naively (at this beginning stage of the book), then, I forecast that EEG use in AugCog will remain stable and increase as it becomes cheaper. BCI use will depend on experimental success but will stay at its current rate of use, barring breakthrough demonstrations of its intended application. BCI use could decrease if AugCog becomes more skeptical of its clinical promise or if it becomes more invasive (more attached to the brain). Either way, BCI innovation will continue over a modest time interval. NNs will continue seeing use recreationally and will be used more interestedly given current successes shown by ChatGPT. By the end of this book, I may be forced to change my mind on all of these counts. We will see how these forecasts pan out after each of the three neurotechnologies has been discussed!

REFERENCES

Deus Ex Wiki, "Augmentation", *Fandom*, n.d. https://deusex.fandom.com/wiki/Deus _Ex_Wiki
Glenn, J. C., Florescu, E., The Millennium Project Team, *State of the Future V. 19.0*, The Millennium Project, 2017.

Hancock, M., et al., Field-Theoretic Modeling Method for Emotional Context in Social Media: Theory and Case Study. In: Schmorrow, D. D., Fidopiastis, C. M. (eds) *Foundations of Augmented Cognition. Lecture Notes in Computer Science*, vol. 9183, Springer, Cham, 2015. https://doi.org/10.1007/978-3-319-20816-9_40

Hancock, M., Nuon, N., Tree, M., Bowles, B., Hadgis, T., A Field Theory for Multidimensional Scaling. In: Schmorrow, D. D., Fidopiastis, C. M. (eds) *Augmented Cognition. Theoretical and Technological Approaches. Lecture Notes in Computer Science*, vol. 12196, Springer, Cham, 2020. https://doi.org/10.1007/978-3-030-50353-6_18

Low, P., Panksepp, J., Reiss, D., Edelman, D., Van Swinderen, B., Koch, C., The Cambridge Declaration on Consciousness, publicly proclaimed in Cambridge, UK, July 7, 2012, Francis Crick Memorial Conference on Consciousness in Human and Non-human Animals.

Minsky, M., Why Freud Was the First Good AI Theorist. In: More, M., Vita-More, N. (eds) *The Transhumanist Reader: Classical and Contemporary Essays on the Science, Technology, and Philosophy of the Human Future*, 2013. https://doi.org/10.1002/9781118555527.ch16

Müller, O., Rotter, S. Neurotechnology: current developments and ethical issues. *Frontiers in Systems Neuroscience*, 11, 2017. https://doi.org/10.3389/fnsys.2017.00093

Schmorrow, D., *Augmented Cognition*, 2014, July 4. https://www.schmorrow.com/

Sood, S., et al., Holarchic Psychoinformatics: A Mathematical Ontology for General and Psychological Realities. In: Schmorrow, D., Fidopiastis, C. (eds) *Augmented Cognition. Lecture Notes in Computer Science*, vol. 11580, Springer, Cham, 2019. https://doi.org/10.1007/978-3-030-22419-6_24

Stanney, K., Winslow, B., Hale, K., Schmorrow, D., Augmented Cognition. In: Boehm-Davis, D. A., Durso, F. T., Lee, J. D. (eds) *APA Handbook of Human Systems Integration*, American Psychological Association, Washington, DC, 2015, 329–342.

The Millennium Project, Real-Time Delphi – The Millennium Project, *The Millennium Project: Global Futures Studies & Research*, 2023. https://www.millennium-project.org/rtd-general/

Electroencephalography (EEG)

2

2.1 FROM NEURAL ACTIVITY TO AUGMENTED COGNITION

How do we get from neurons to thought? There are at least three possibilities. One is that the nervous system and mind are identical. More specifically, the central nervous system (CNS), which consists of the brain (along with the spinal cord) would be identical with consciousness. The next possibility is that the brain enables the mind: without the brain, there would be no mind. This is a causal thesis that a biological entity (the brain) gives rise to a psychological one (the mind). Thirdly, the mind and brain are distinct, but they intersect at least enough to allow for an exploitable relation between them. This third possibility is perhaps the most tenable as far as neurotechnological augmented cognition goes.

It is also possible that the mind and brain have no relation. If this were true, cognitive neuroscience would be somewhat less intuitive as an interdiscipline. If cognition and neurology had no relation, what would be the point of studying them together? The only real answer to this could be pure curiosity.

What can the activity of neurons do to better our thoughts? Some—perhaps most—believe neurons are, indeed, the basic unit of thought. I do not necessarily conceive of them this way. It is still too early (at best) to admit such a claim about reality that cuts across two distinct research areas, viz. those of cognition and neurology. How could we tell if a neuron corresponded with a cognitive unit? The brain can be altered in ethically proper ways for consensual research,

DOI: 10.1201/9781032692982-3

FIGURE 2.1 Portable EEG worn by anonymized subject

and cognition could be measured afterward, but some aspects of cognition may remain obscure to researchers operating from a primarily physicalist (and non-mentalist) point of view. But if EEG can augment cognition, it is likely to augment a researcher's understanding of an EEG helmet-wearer's cognitive state (e.g., conscious).[1]

As a psychologist, I consider cognition purer a topic for psychology than I do neurology. Both cognition and neurology play large roles in psychological research, but when neural studies do, they always make a study neuropsychological, neurological, or neuroscientific. This book is not intended to be a philosophical study of the nature of reality, but some of my augmented cognition research is concerned with ontology (and—in a somewhat different sense than I have treated ontology—so is computer science). It is worth being clear about the subjects discussed.

It is well-known that electroencephalography (EEG) affords scans of broad brain activity. It is useful for studying regional rather than individual neural activity (where the latter is more the province of functional magnetic resonance imaging, i.e., fMRI). EEG has been used to study personality (see, e.g., Nardi, 2011). EEG consists of an electrode headset that is attached to the participant's scalp via cool, wet gel. EEG is of interest to augmented cognition insofar as it allows us to learn what different, very broad brain regions are up to, especially during tasks. These tasks vary in nature.

The group NeuroTechX has recently written a helpful primer for EEG (and BCI). EEG became accessible to the open-source community during the 1990s. Experimenter Hans Berger is credited with having invented EEG and coining its name in 1929. EEG is "a technique with over a hundred years of history, and while...originally used more strictly in the fields of psychology, medicine, and neuroscience...is widely used today in gaming, *human–computerinteraction* [emphasis added], neuromarketing, simulations, and beyond"

---

[1] A further consideration involved with EEG hardware includes properly placing the helmet onto the subject's head. Precision is important since the brain region activity measured depends on sensor placement.

(Farnsworth, November 20th, 2023). Revlin (2013) noted that EEG "is the oldest imaging method of all those currently employed" (p. 35).

EEG is described as being used to make relative measurements, in that electrical currents picked up are contrasted with a frame of reference. This frame is to be a place absent of the anticipated neural action. NeuroTechX (2023) stated further that "it is important to verify axis labels when viewing EEG graphs" (p. 12). Measured action could be spontaneous, evoked, or induced. EEG is largely utilized for researching the neural structure of cognition. NeuroTechX wrote that ERP is among the more utilizable functions of measurement via EEG. EEG offers lower spatial resolution than other brain-imaging technologies but good temporal resolution: EEG can pick up neural firing with a precision of one-thousandth of a second (1 millisecond). Studying electrical currents with EEG takes place in two phases: data acquisition and analysis. Acquiring data via EEG includes electrodes (which pick up signals), amplifiers (that "view and interpret" signals), and a computer (that consolidates signals) (p. 15).

Wet-gel electrodes have been described. NeuroTechX also noted that "dry electrodes", which "leverage…flexible material, a stable structure, dry conductive surface…and a sophisticated algorithm" (p. 17) can be used. The algorithm is used to pull and detach signals via EEG from bio-environmental sources of static. Thus, dry electrodes could be used in lieu of wet ones to make the process of data acquisition cleaner. NeuroTechX also distinguished between *active* and *passive* electrodes. Active ones allow signals to be amplified right at the site of measurement (i.e., the top of the head), all prior to making it to the ordinary amplifier setup. Passive electrodes are usually used, as they are more cost effective. Amplifiers are used since EEG signals are small, measurable in millivolt, microvolt, or nanovolt units (mV, μV, and nV, respectively). Following amplification, signals become digitally rendered, such that they become processable and then viewable via computational monitor.

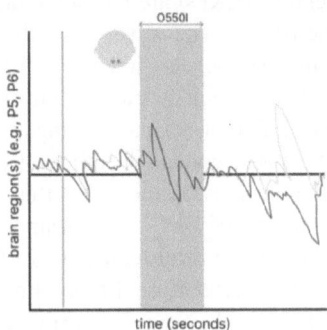

FIGURE 2.2 An example of an ERP reading

EEG signals are recorded via computers. Such computers could be either desktops or laptops—provided, NeuroTechX noted, that the computer in question possesses enough capability and agility to stay abreast of the huge volume of data fed via amplification. According to NeuroTechX, the majority of state-of-the-art EEG labs usually have two or three such laptops or desktops. Computer #1 shows, if experimentally necessary, stimulus usually in the form of either an audio or eye task; computer #2 is the signal recorder. It is crucial that computer #2 knows which stimulus it is being shown, and when. The possibility also exists, however, that recording and stimulus be synched via "time stamps" (p. 18). Programs are used to record data, storing EEG signaling. EEG signal analysis needs a certain kind of program to *preprocess* (a term familiar to any readers involved in machine learning) said signals, prior to any alteration of them. These programs exist in both industrial and open-source forms. For a comprehensive treatment of EEG signals, the reader is referred to Siuly, Li, and Zhang (2016).

EEG is used in augmented cognition research for eye-tracking (ET) (e.g., Xiang & Abdelmonsef, 2022) and classification (Wang & Wang, 2022; Rajabi et al., 2023). Rajabi et al. described that their EEG classifier was used "to perform binary classification on the EEG signals collected during the presentation of...face image[s] to the user to distinguish the images that attracted their attention" (Schmorrow & Fidopiastis, 2023, p. 33). Their classifier was "trained for each subject separately using their EEG data collected while performing a target face detection task". Rajabi et al. expected such classifying "to reveal images relevant to the target face from the subjects' point of view".

What makes EEG a human–computer interactive (HCI) technology is the use of software to analyze electrode inputs. A typical number of electrodes to use is between 12 and 64 (Willingham, 2007) to measure the "summed activity of millions of neurons" (p. 54). Different kinds of neurons fire at different rates, but even while resting, they fire. To control for neurons' constant firing, event-related potentials (ERPs) are measured, and "tens or hundreds" of similar trials are done (p. 55). The waves from these trials are averaged, smoothing them out. In my experience—having my EEG taken during a linguistic vocalization and identification task at the University of California, Los Angeles (UCLA)—I noticed Python scripts running at the beginning and end of my participation. I did not get a good look at the code. (This EEG-BCI intersection will be discussed more, close to the end of this book.)

For a better understanding of EEG, the research of LaRocco, Le, and Paeng (2020) is helpful. They list several "consumer EEG-based devices" such as "Neurosky MindWave, InteraXon Muse, Emotiv Epoc, Emotiv Insight, and OpenBCI". In LaRocco et al.'s study, each of these devices was listed as having reportedly enabled detection of drowsiness in people. This demonstrates the variety of consumer EEG devices that have been available to researchers and a focus on a specific application of them. This overlaps with a small cluster of AugCog research that focuses on the mental state of aircraft pilots, including their performance (Russo, Kendall, Johnson, Sing, Escolas, Santiago, Holland, Hall, & Redmond, 2005) and phenomenology.[2] EEG has been "used within augmented cognition systems [to] form situation awareness advisory tools that are able to provide real-time feedback to air-traffic control supervisors and planners" (Abbass, Tang, Amin, Ellejmi, & Kirby, 2014).

2.2 MEASURING INDIVIDUAL COGNITIVE DIFFERENCES

EEG answers the questions of where and when brain events happen. In cognitive research, it could inform an experimenter of when the brain registers stimuli. EEG has also been used in personality research. The benefit of this is to understand how the brains of different individuals work. Dario Nardi, a UCLA researcher, conducted a pilot EEG test to validate the popular Myers-Briggs Type Indicator® neuroscientifically. He was interested in finding distinct regional patterns in participants working on a variety of tasks, noting scattered versus whole-brain readings. Based on these, he classified each of the 16 Myers-Briggs personality types into distinct cognitive profiles based on existing theory in the area.

Personality research constitutes a small minority of augmented cognition research. Work like Nardi's, while not augmented cognition proper, could be thought of as an "extended" augmented cognitive study. I mean this in the sense that the reader leaves with an enriched understanding of cognitive neuroscience in a more neutral psychological area (as opposed to a clinical context—more so the province of BCI research). Nardi's findings were published in a more popularly geared book, *The Neuroscience of Personality*. This study to my mind is exemplary of how interesting EEG research can be, despite its status as the most popular (and accessible, compared to BCI) brain technology of those discussed here. Nardi's work shows how EEG can be used to understand personality at a more base level.

Cervera-Torres, Minissi, Greco, Callara, Ferdowsi, Citi, Maddalon, Giglioli, and Alcañiz (2023) proposed a model including socio-emotional virtual reality (VR) and measurement of emotional response. The latter includes

---

[2] Also worth exploring is Berka, Levendowski, Davis, Whitmoyer, Hale, and Fuchs' (2006) research on situational awareness and EEG (including "time-locked potentials…generated by neuronal networks").

EEG, which the authors proposed would be used to record emotional response "during the social-emotional and non-social VEs [virtual environments]" (Schmorrow & Fidopiastis, 2023, p. 323). Cervera-Torres et al.'s study consisted of a proposal to use EEG in this manner, as well as to study "emotional HBO [human body odor]" and affective valence (ranging from negative to positive). Their study is indeed unique within AugCog: not only for its hypothetical nature but also given the authors' interests in VR, body odor, and affect. The study also discusses social cognition and behavioral cues, making it more holistically psycho-neurotechnological. It is conceivable that this kind of study, if carried out, would open the door to more studies involving VR and our sensory apparatus. It is widely known that a limitation of current VR is its inability to elicit the faculties of taste or smell, and studies like these could eventually change this.

2.3 EEG IN GENERAL

EEG in general is used clinically to troubleshoot neurological issues like injury and abnormalities, including tumors and epilepsy. A good primer on the history of EEG in general is Giannitrapani and Liberson's (1985) book, available freely online. This book covers a good range of kinds of EEG studies focusing on mental ability. Also included are (regional) spectral analysis, brain function by area, factor analysis, and the function and age relativity of EEG frequency, as well as EEG hardware, conditions, and recordings. Skills needed to learn how to use EEG include properly placing the helmet, managing the signal-to-noise ratio (including noticing noise and lowering it), preprocessing of data (at least, in an EEG-based BCI system), and software engineering.

I have, perhaps, taken for granted up until this point that the reader of this book knows about EEG at a basic level. This includes EEG's function, or what it does. In her book *Quirk: Brain Science Makes Sense of Your Peculiar Personality*, author Hannah Holmes stated this thus: "EEG measures electrical changes in the brain" (p. 22). More technically, EEG is used for ERPs. The unit of EEG measurement of brain activity is the microvolt (μV), and the upper bound for the interested frequency of waves is around 30 Hertz (Hz) (Smith). Amplitude is the other dimension of measurement, showing the height of recorded brain waves. Brain waves show the "total electrical output of neurons near…[EEG] electrodes" (Revlin, 2013, p. 36), depicting "continuous brain activity" (Willingham, 2007, p. 54). NeuroTechX is more specific, having written that EEG allows for the recording of the brain cell action of over a thousand neurons as they fire in concert with one another: this allows for the measurement of broader, regional brain activity.

Holmes' is the most elegant framing I have come across. Yet she went further, noting that EEG measurement is conducted when people talk, solve mathematical problems, and view photography. Each of these is of interest for augmented cognition: discourse analysis could be augmented; mathematics is frequently taken up (sometimes innovatively, and often to great experimental effect) in augmented cognition; aesthetics is present in user interface (UI) work, which is a prominent part of the face, culture, and work of HCI. Based on our current understanding of the brain regions and their functions, hypothesizing which are active during a given task should be reasonably easy.

The eminent theoretical physicist Michio Kaku discussed neurotechnology at some length in his popular book, *The Future of the Mind* (2014). He noted that EEG dates to 1924; only in recent times, though, could the computer be used to make comprehendible what the EEG measures through electrodes. Cutting-edge EEG, Kaku noted, consists of placing a hairnet with miniscule electrodes on top of the scalp. He described EEG as "strictly passive" (p. 26), in that with it, miniscule electromagnetism traverses the brain. This enables measurement of whole-brain patterns taking place as people sleep, focus, dream, and relax. Usually, EEG imaging shows the emission of gradual EMG currents when participants are awake. Kaku's most simple statement about EEG is that such imaging measures electric brain currents immediately. NeuroTechX (2023) wrote that EEG records such currents flowing as

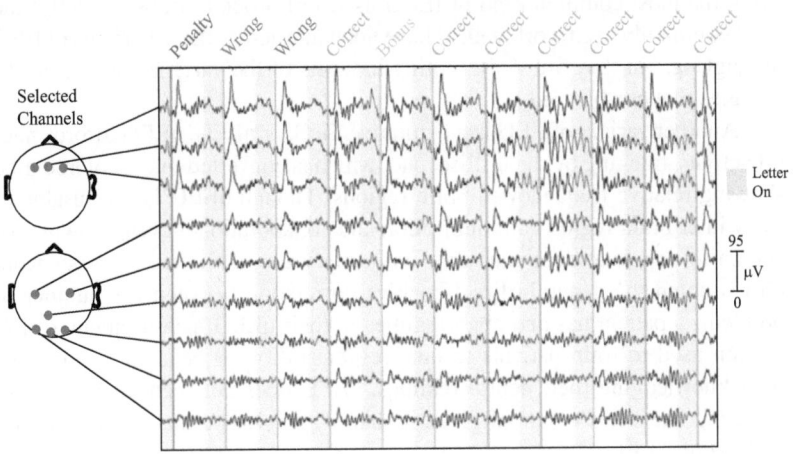

FIGURE 2.3 An example of an EEG readout outside of AugCog (University of California, San Diego, 2023)

the pyramidal neuronal layer is excited. EEG's largest boons (in addition to temporal resolution) are being both cheap and convenient.

Kaku also discussed EEG helmets and typewriters. Helmets are the most common implementation of EEG. A benefit of EEG typewriters, wrote Kaku, is their noninvasive nature (as compared with electrocorticogram, i.e., ECOG). EEG typewriters trade off precision and accuracy for being relatively easily accessible. The Austrian company Guger Technologies is an example of a company that has seemed to work successfully with this new kind of EEG.

Kaku introduced work done with magnetoencephalography (MEG). Kaku wrote that MEG could have complemented EEG sensor work. However, "true telepathy helmets" were still "many years away" (p. 71). He also provided a possible answer to the *binding problem* in memory research. Concerned with how our minds consolidate memories into singular experiences, Kaku proposed the partial answer through "the fact that there are electromagnetic vibrations oscillating across the entire brain at roughly forty cycles per second": such vibrations could be sensed via an EEG scan (p. 107).

A variant of EEG known as iEEG, or intracranial EEG, exists. Use of iEEG is more invasive than standard EEG; instead of electrodes being placed on the scalp, they are surgically inserted into the skull. iEEG's functions are twofold: it records electrical impulses from the cerebral cortex and stimulates the brain (NeuroTechX, 2023). iEEG leverages *neuromodulation*, targeting specific parts of the brain electrically or pharmaceutically to record signals and stimulate the brain. Neuromodulation is done for both practical and non-applied purposes. Another variant of EEG is sEEG, or stereoencephalography, which is actually a form of iEEG. NeuroTechX noted that sEEG went from being the most common kind of iEEG used in Europe to (most recently) the most commonly used worldwide. I have not come across these variants of EEG in AugCog, but they are worth noting in terms of the varieties of neurotechnologies that can be used.

Another variant of EEG is quantitative EEG, or qEEG. qEEG affords neural activity-tracking by the millisecond, which is converted into a colored map showing relative frequency in brain regions. Though qEEG is also usable to show inter-brain region communication and collaboration—even allowing for person-to-population comparison—its primary use is clinical. Van der Kolk (2015) noted the general, direct positive correlation between the number of problems a patient has and abnormalities in their qEEG measurement. qEEG is even used to train patients to take responsibility for behavioral problems by acknowledging their neural realities. qEEG does not seem to be used in AugCog, but it has potential given it shares inexpensiveness and portability with regular EEG.

It is debatable whether the study of emotion fits into augmented cognition. Some of my augmented cognition research has advocated for an expansion of topics to include augmented affect (the unconscious side to emotion). Certainly, both emotion and cognition are psychological topics: further, healthy affect lends itself to better-quality cognition (and perhaps vice versa). Emotion, or technically affect, is usually treated as part of cognitive psychology (though it is studied in other disciplines, like philosophy). Richard J. Davidson discusses at length in his book, *The Emotional Life of Your Brain* EEG in relation to emotion. The book offers a good history of EEG from a seasoned EEG researcher.

Less debatable for AugCog could be the use of EEG to study the brain's default mode network (DMN). The DMN has become one of the most prominent regions of study in neuroscience. Knyazev, Slobodskoj-Plusnin, Bocharov, and Pylkova (2011) studied the DMN using EEG, noting that positron emission tomography (PET) and functional magnetic resonance imaging (fMRI) were more common and less controversial options for such. However, their work opens the door to AugCog researchers studying the DMN: this is especially relevant given the breadth of DMN's involvement in cognition: it is involved in "self-referential processing, interoception, autobiographical memory retrieval, [and] imagining [the] future" ("Default Mode Network", 2011).

EEG is also used in neurofeedback (e.g., "Who We Are I Brain Performance"). Brain Performance—a group of neurofeedback centers—advertises that they process "EEG brain maps" with a "research-based EEG normative database"; their methodology is "guided by neuroimaging". This company's goal is to enhance physiological processes responsible for maladaptive physical or behavioral symptoms. Though they do not seem as interested in augmenting cognition directly, treating clients' physiologies can lead to alleviation of maladaptive symptoms that block optimal cognitive functioning.

Within HCI, EEG enjoys broader use than just in AugCog. It has also been studied within engineering psychology and cognitive ergonomics, artificial intelligence (AI), and in the conference proceedings titled *HCI in Business, Government, and Organizations* (2023). Potentially relevant for augmented cognition—certainly so for the closely related area of positive computing—is Davidson's having measured prefrontal cortex (PFC) activity in relation to resilience (capital-R Resilience, for him). Resilience is here conceptualized in relation to positive and negative emotion and associated with "left-right asymmetry" (p. 202) in the PFC. Davidson also used EEG to study expert meditation (his participants included the famous Matthieu Ricard). He found that the brain's gamma patterns were stronger for participants while they meditated: so strong that they set a scientific record. Davidson also found that these gamma patterns endured after meditation for the eight monks, showing stable brain changes resulting from regular and disciplined meditation.

EEGs consist of a helmet used to record signals from a person's brain. Neural activity tracked this way is first cleaned, then analyzed using software.

The analysis of cleaned data consists of noting patterns in neural activity. Said patterns may be active while a participant carries out a task (usually while sitting, but not necessarily). Alternately, patterns of brain activity may be compared between individuals who are part of the same study.

REFERENCES

Abbass, H. A., Tang, J., Amin, R., Ellejmi, M., Kirby, S., Augmented Cognition using Real-time EEG-based Adaptive Strategies for Air Traffic Control, 2014. https://www.husseinabbass.net/papers/HFES2014Abbass.pdf

Berka, C., Levendowski, D. J., Davis, G., Whitmoyer, M., Hale, K., Fuchs, S., Objective Measures of Situational Awareness Using Neurophysiology Technology, Augmented Cognition Conference, 2006. https://www.researchgate.net/publication/236610901_Objective_Measures_of_Situational_Awareness_Using_Neurophysiology_Technology

Cerveras-Torres, S., Minissi, M. E., Greco, A., Callara, A., Ferdowsi, S., Citi, L., Maddalon, L., Giglioli, I. A. C., Alcañiz, M., Modulating Virtual Affective Elicitation by Human Body Odors: Advancing Research on Social Signal Processing in Virtual Reality. In: Schmorrow, D. D., Fidopiastis, C. M. (eds) *Augmented Cognition. Lecture Notes in Computer Science*, vol. 14019, Springer, Cham. https://doi.org/10.1007/978-3-031-35017-7_20

Davidson, R. J., *The Emotional Life of Your Brain: How Its Unique Patterns Affect the Way You Think, Feel, and Live—and How You Can Change Them*, Hudson Street Press, New York, 2012.

Default Mode Network, *The Neuroscience of Depression*, Default Mode Network - an overview | ScienceDirect Topics, 2021.

Farnsworth, B., "EEG Headset Prices – An Overview of 15+ EEG Devices", iMotions, November 20, 2023. https://imotions.com/blog/learning/product-guides/eeg-headset-prices/

Giannitrapani, D., Liberson, W. T., *The Electrophysiology of Intellectual Functions*, Karger, New York, 1985. https://archive.org/details/electrophysiolog00gian/page/n6/mode/1up

Holmes, H., *Quirk: Brain Science Makes Sense of Your Peculiar Personality*, Random House, New York, 2011.

Kaku, M., *The Future of the Mind*, Doubleday, New York, 2014.

Knyazev, G. G., Slobodskoj-Plusnin, J. Y., Bocharov, A. V., Pylkova, L. V., The Default Mode Network and EEG Alpha Oscillations: An Independent Component Analysis, *Brain Research*, 1402, 2011. https://doi.org/10.1016/j.brainres.2011.05.052

LaRocco, J., Le, M. D., Paeng, D. G., A Systemic Review of Available Low-Cost EEG Headsets Used for Drowsiness Detection, *Frontiers in Neuroinformatics*, 14, 553352, 2020. https://doi.org/10.3389/fninf.2020.553352 PMID: 33178004; PMCID: PMC7593569.

Nah, F., Siau, K., *HCI in Business, Government and Organizations (Part II)*, Springer, 2023.

Nardi, D., *Neuroscience of Personality*, Radiance House, 2011.

NeuroTechX, NeuroTechX – Slack Group for Neuroscience, Hive Index, 2023. https://thehiveindex.com/communities/neurotechx/

NeuroTechX, *The Neurotech Primer: A Beginner's Guide to Everything Neurotechnology*, 2023.

Rajabi, N., Chernik, C., Reichlin, A., Taleb, F., Vasco, M., Ghadirzadeh, A., Björkman, M., Kragic, D., Mental Face Image Retrieval Based on a Closed-Loop Brain-Computer Interface. In: Schmorrow, D. D., Fidopiastis, C. M. (eds) *Augmented Cognition. Lecture Notes in Artificial Intelligence*, vol. 14019, Springer, Cham, 2023, 26–45.

Revlin, R., *Cognition: Theory and Practice*, Worth Publishers, New York, NY, 2013.

Russo, M., Kendall, A., Johnson, D., Sing, H., Escolas, S., Santiago, S., Holland, D., Hall, S., Redmond, D., Relationships Among Visual Perception, Psychomotor Performance, and Complex Motor Performance in Military Pilots During an Overnight Air-refueling Simulated Flight: Implications for Automated Cognitive Workload Reduction Systems. In: Schmorrow, D. (ed) *Foundations of Augmented Cognition: Volume II*, Lawrence Erlbaum Associates Publishers, Mahwah, New Jersey, 2005, 174–183.

Schmorrow, D. D., Fidopiastis, C. M. (eds), *Augmented Cognition*, Springer, Cham, 2023.

Smith, E. J., EBME & Clinical Engineering Articles: Introduction to EEG, *ebme*. https://www.ebme.co.uk/articles/clinical-engineering/introduction-to-eeg

Siuly, S., Li, Y., Zhang, Y., *EEG Signal Analysis and Classification: Techniques and Applications*, Springer, 2016. L-8486706-e230c729d9.pdf (e-bookshelf.de)

University of California, San Diego, EEG / ERP data available for free public download, updated 2023. https://sccn.ucsd.edu/~arno/fam2data/publicly_available_EEG_data.html

Van der Kolk, B., *The Body Keeps Score: Brain, Mind, and Body in the Healing of Trauma*, Penguin Books, New York, 2015.

Wang, X., Wang, Z., CNN with self-attention in EEG classification In: Kurosu, M., et al. *HCI International 2022 - Late Breaking Papers. Multimodality in Advanced Interaction Environments*. HCII 2022. Lecture Notes in Computer Science, vol 13519. Springer, Cham, 2022. https://doi.org/10.1007/978-3-031-17618-0_36

Who We Are | Brain Performance, *Brain Performance: Neurofeedback Centers*, n.d. https://www.brainperformance.com/about-us

Willingham, D. T., *Cognition: The thinking Animal*, 3rd edn., Pearson Prentice Hall, New Jersey, 2007.

Xiang, B., Abdelrahman, A., Vector-based data improves left-right eye-tracking classifier performance after a covariate distributional shift, 2022. https://doi.org/10.1007/978-3-031-17615-9_44

Neural Network (NN) 3

3.1 BRAIN SIMULATION TO AUGMENT COGNITION

The brain is a neural network. There are at least as many such biological NNs as there are brains in the world. Complexifying this even further, a single brain could be thought of as a network of neural networks (perhaps a neural network-network), or a network of lobes that are themselves neural networks. A network of brains could also be considered such a meta-neural network: albeit, one that is intra-connected by interfacing brained beings who interact via their outer bodies and their projections. As can already be seen here, the concept of the neural network enjoys some flexibility even before one considers artificial NNs (like those that much AugCog research favors).

The human body is heavily neural-networked. This claim is more nominal and analytical than ontological. What is meant here is that neural networks are diverse in type and levels of analysis. We call different kinds of things neural networks, ranging from biological to technological systems. Within the body, one can zoom out from a basic neural network of a cluster of communicating neurons to the entire nervous system. It is true in reality that the human body is heavily neural-networked, but this statement is multidimensional.

As was done with augmented cognition, an understanding of each part of the neural network concept will be undertaken. "Neural" refers to something sufficiently resembling, if not outright being, nerve cells. Nerve cells exist in organisms' bodies and are simulated in highly simplified form via NNs. Networks are compositions of interrelated nodes; they may be social, technological, or biological (Boyd & Batchelder, 2016, pp. 413–414).

We simulate real-world systems in part to make them tractable. Alternatively, we may perform experiments involving simulations (as my augmented cognition team and I have done, in a different context) with less worry

DOI: 10.1201/9781032692982-4

about the ethics of human-subjects research. Such research often requires approval by an institutional board, such as the Institutional Review Board (IRB). Though brain experimentation has a checkered history—viz., lobotomies and lesions conducted with no knowledge of what negative outcomes could result—it is now done often and unproblematically. In any case, neural nets are a way to sidestep human concerns, while also having potential benefits for humans (if findings are extendible to their lives) and being interesting technology. Neural networks are fascinating due to their close tie to neurocognitive modeling.

The pioneering artificial intelligence researcher Marvin Minsky wrote of "neural-net research" in 1963 (p. 303). His project at the Massachusetts Institute of Technology (MIT) was titled "Neural Nets and Theories of Memory". He wrote on how neural networks could act as "learning machines" (p. 303). More generally, he wrote on the ebbs and flows that NN research during his time had taken, ranging from ambitious to focusing on simple modeling and problem-solving. Minsky's analysis includes past NN research trends spanning across a decade, the beginning of the field, and current (at his time of writing) trends.

Neurons are useful for neural nets (NNs) since the former are "all-or-nothing". That is, they either fire or do not, making them binary. It is thus easy to imagine how they are computerized as simple logic gates: 0 equals "off" (do not fire); 1 means "on" (fire). Neurons are elaborate message-senders, where their messages are neurotransmitters with certain functions. (There is, however, apparently work to be done making neurotransmitters amenable to the neural net approach.) Neural net neurons are thus representable in binary code, though they are not usually used toward this end. Instead, such neurons exist as nodes of a broader "graph" that is the brain. Connections between nodes are represented by graph-theoretic edges, metaphorical synapses (gaps between neurons across which neurotransmitters travel).

A backbone concept within NN (as well as in machine learning and artificial intelligence) is that of the *perceptron*. The Rosenblatt-variant perceptron has been described as a basic NN made up of one, humanmade neuron, and as being "the basis of today's complex neural networks" (Neuroelectrics, 2023, October 3). Perceptrons are depicted visually in terms of an input layer, a "hidden layer" in the middle, an output layer, and possibly also a result. The basic mathematics of a perceptron is useful for rendering it more intuitive. After data is fed to it via the input layer, a summative processing function would take the form of:

$$Sum = \Sigma \ (Input \times Weight) \qquad (3.1.1)$$

Input values are multiplied by weights assigned to variables. This result is summed for all input and weight values, yielding a sum that is next compared to the activation threshold in the output layer. If the sum exceeds the activation threshold, the simulated neuron perceptron "fires".

Multilayer perceptrons (MLPs) in the above sense are synonymous with "vanilla neural networks" (baeldung, 2023, June 13) (VNNs). However, such perceptrons are distinguishable from deep NNs (DNNs). DNNs have comparatively more hidden layers, take longer to train, and can operate using a tensor processing unit (TPU) (rather than a graphics processing unit—i.e., GPU—which suffices for an MLP). Both MLPs and DNNs are types of neural networks, with MLPs having three or more layers. DNNs are utilized "by social networks and search engines for photo tagging" (Corella & Lewison, 2019, p. 42), wherein these NNs "match or surpass" human performance.[1] The pyramidal neuron is an example of a two-layer NN (Poirazi, Brannon, & Mel, 2003).

The hidden layer of the perceptron illustrates the concept of the *black box*. In psychology, the black box refers to the belief held by behaviorists that cognitive processes are unknowable. Since the cognitive revolution from the 1970s onward, this belief has been disposed of, and most modern introductory psychology textbooks define the discipline as being concerned with behavior and mental processes. Much theory and experimentation have since been invested in uncovering these processes. Freudian theory was a good start to this for Western psychology, as was the earlier introspectionism (which many came to dismiss given its unfalsifiability). For AugCog, the most relevant part of this has been attempting to describe cognition via the computer metaphor. To what extent can cognitive processes be analogized to computation?

Despite its ubiquity and variety, the utility of the neural net could not be overstated for the current millennium. It is perhaps the most interdisciplinary technology focused on in this book, given that it can be studied in bioinformatics, neurotechnology, computational neuroscience, and computer science (in addition—of course—to its roots in mathematics and neurology). If EEG is the most widespread and well-known of the three neurotechnologies discussed in this book, NN may be the broadest in this multidisciplinary sense. NN thus has great potential to catalyze interdisciplinary research, an underappreciated fact in the current century of the brain.

NN is also closest to another technology relevant for augmented cognition: artificial intelligence (AI). AI naturally deserves its own space to be discussed: suffice it to say that any connection to it will make a technology more visible to the public. NN is undoubtedly more niche and less well known; some might even opt to treat it as a subset of AI. I believe this is not necessarily a suitable position in augmented cognition. Intelligence is too broad to study in just the cognitive way (preferring a more holistically psychological framing).

---

[1] NNs—specifically, deep Q-learning ones—are even taught to students so they can "land a virtual lunar lander on Mars" (DeepLearning.AI, 2024)!

Thus, if cognition and brain are one, intelligence cannot be totally reduced to either. NN makes no claim of being a cognitive technology, even though it may turn out to be directly relevant to it. NN is a much more modest, yet still highly useful, brain technology.

One last thing to note about NN nodes is that they are basic input-output systems (BIOSs). They do not pretend to be as sophisticated as biological neurons: they lack myelin sheaths (which facilitate movement along the neuron), axons, and dendrites. One could certainly scale up an NN to include such features, but there are benefits to the simplistic node discussed thus far. It is too early to say what type of NN would be best for quantum computing. It is likely that neurons here would be best modeled more precisely, but this ventures into the area of neurophysics. Usually, the "process" aspect of the NN node's input-process-output structure is of interest, though it may also be treated as a black box (at least from the outside).

How does an engineer use brain technologies like the NN? Some develop NNs as pet projects, sharing their thoughts and progress in online spaces like forum boards, or "virtual communities of practice" (VCoP). I have learned much about the process of building NNs simply by perusing forum boards through which others have shared their experiences. Through this, they inform one another of the specific technologies they use and build a sense of techie solidarity.[2] While NNs thus have the power to bring developers together, they also have the power to potentially substitute for an organism's brain. These are arguably the NN's two most powerful reasons for being.

How else do neural nets augment cognition? Perhaps their most used way of doing so is image recognition. Neural nets are trained with data that allow them to predict things like who or what is in a photograph. A "deep" NN utilized to verify faces in images is "trained with millions of labeled faces belonging to thousands of people.... [Its] outputs are...compared according to some similarity metric and deemed to belong to the same person if their similarity metric is above a certain threshold" (Corella & Lewison, 2019, p. 32). Corella and Lewison noted that in Google's FaceNet, NN "output is a vector with 128 coordinates, each of which is a single byte, and the similarity metric used to compare the vectors derived from...verification samples is the Euclidean distance between the two vectors". Face verification had been "disrupted" "by...the advent of deep learning (p. 41). An area for future research is cybersecurity in that deep learning was found to be "lacking in protection against presentation attacks".

---

[2] This was observed in a VCoP by participants I consider in an immediately peripheral group to AugCog. Such technologists have shown some interest in AugCog but prioritize their tech work and hobbies over it.

3.2 ARTIFICIAL VERSUS BIOLOGICAL NEURAL NETWORKS (ANNS VS. BNNS)

A neural network is a web of neurons. Said neurons are abstract and metaphorical (artificial) or biological. They are connected via line segments (edges) in ANNs and are separated by synapses in biological brains. Brains are complex structures made up of between 80 and 100 billion neurons. Given this, it may be impractical to model the entire brain at the neural level artificially. Indeed, while brains are made up of tens of billions of units, artificial NNs are vastly simpler (and are structurally different in terms of their implementation).

Neural networks come in "A" (artificial), "B" (biological), "C" (convolutional), "D" (deep, or dynamic as in Deng, Xu, Wang, Wang, and Chen, 2016), "DC" (deep convolutional), "DS" (deep spiking), "E" (ensemble), modular, "P" (probabilistic), "R" (recurrent), "RBF" (radial basis function), "(BT-)Rv" (Beam Tree Recursive) (Chowdhury & Caragea, 2023a, 2023b[3]), "S" (Tavanaei, Ghodrati, Kheradpisheh, Masquelier, & Maida, 2019), and "V" variants, making the neural net the most variable brain technology discussed.[4] It is also the most detached, least embodied, and least hard in terms of being disproportionately implementable via software (rather than hardware, like a head helmet).

Artificial NNs (ANNs) are relevant to other kinds of NN: for example, Tavanaei et al. noted that spiking NNs (SNNs) are "more biologically realistic than ANNs", going as far as to write that they may be "the only viable option if one wants to understand how the brain computes" (p. 1).[5] Unlike EEG and BCI, NNs are not worn—they are only written and implemented in software. Artificial neural nets (ANNs) have already been discussed earlier in this book (if somewhat indirectly). The view exists that neural nets are a subset of AI (see Giacomelli's work, for example).

It is a simple (but perhaps not trivial) fact that both artificial and biological NNs run at least partially on electricity. The distinction between ANNs and biological neural nets (BNNs) is an easy one—the former are virtual models of the latter. Only the latter occur in nature prior to human invention. The former

---

[3] BT-RvNN—i.e., Beam Tree Recursive NN—was represented as a "sentence encoder" and a "token contextualizer".

[4] NNs can even be organizational. The internet could be conceptualized as an abstract, human neural network. This idea is implemented by the nonprofit think tank The Millennium Project, which conducts future studies and has "Nodes" of relatively centralized activity around the world.

[5] However, these authors also conceded that SNNs are not as accurate as ANNs. Trend-wise, they stated that this "gap is decreasing".

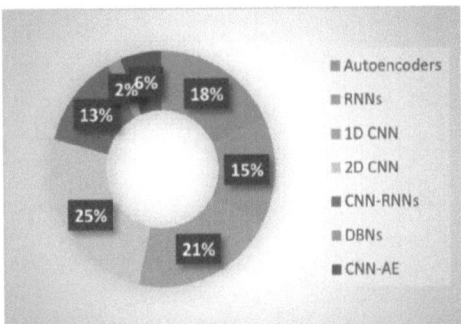

FIGURE 3.1 Percentages of deep learning techniques reviewed in Saminu, Xu, Zhang, Kader, Aliyu, Jabire, Ahmed, and Adamu (2022). From beginning to end, the percentage wheel starts with Autoencoders (18%) and ends with CNN-AE (6%). © Saminu et al. 2022, CC BY 4.0.

might be more amenable to change, assuming our species is done evolving (and that our brains have evolved gradually over time, adding new layers with the neocortex being the most recent). There is no real need to labor over this distinction: suffice it to say that we can work more freely modifying how ANNs work versus BNNs. BNNs are more immediately impacted by human concerns like injury; ANNs are for our understanding and potential augmentation. The brain is a complex web of neurons interacting with one another via neurotransmitter signals.

ANNs can be built using various open online resources. There are neural network libraries available using multiple programming languages to help build neural nets. JavaScript libraries for building neural nets include BrainJS—utilizable in both web browsers and the backend Node.js framework—and Synaptic, with which one can create "recurrent and second-order networks" (Kundariya, 2022, September 27). Other languages one can create neural nets with include Python and C++. (Programs for such can range from one line to 600!) One of the ways ANNs can augment cognition is by allowing us to simulate population growth.

It is probably of interest to current followers of AI that ChatGPT, created by OpenAI, is based on "neural network architecture" (Cretu, 2023, July 21). ChatGPT is a large-language model (LLM): it is a kind of AI algorithm which uses neural net methods and parameters to interpret language via supervised learning. Thus, neural networks come in various architectures and methodologies.[6] ChatGPT is prompt based—that is, it learns based on prompts input-

---

[6] Some research using neurotechnology has been motivated by the goal of elucidating cognitive architecture. UCSB cognitive neuroscientist Scott Grafton's past work used fMRI, presuming that a cognitive architecture underlies goal-directed behavior. The thrust of this work is that

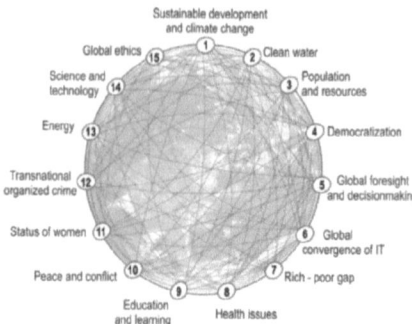

FIGURE 3.2 "Global Challenges" (Glenn, Florescu, and The Millennium Project Team, 2019, p. 7). The Millennium Project's depiction of global challenges is represented using nodes and edges, representing a metaphorical neural network.

```
import numpy, random, os
lr = 1 floating rate
bias = 1 #value of bias
weights = [random.random(),random.random(),random.random()] #weights generated in a list (3 weights in total for 2 neurons and the bias)
#define iteration of neurons and ...
def Perceptron(input1, input2, output) :
    outputP = input1*weights[0]+input2*weights[1]+bias*weights[2]
    if outputP > 0 : #activation function (here Heaviside)
        outputP = 1
    else :
        outputP = 0
    error = output - outputP
    weights[0] += error * input1 * lr
    weights[1] += error * input2 * lr
    weights[2] += error * bias * lr
#create first defining output neuron's work
''' Undo parameters (neurons' two values and expected output)
    outputP variable corresponds to Perceptron's output'''
    #calculated error, to modify weights of every connection to output neuron right after'''
for i in range(50) :
    Perceptron(1,1,1) #true on true
    Perceptron(1,0,1) #true on false
    Perceptron(0,1,1) #true on true
    Perceptron(0,0,0) #false on false
#loop makes AN iterate each t (learning)
x = int(input())
y = int(input())
outputP = x*weights[0] + y*weights[1] + bias*weights[2]
if outputP > 0 : #activation function
    outputP = 1
else :
    outputP = 0
print("x", y, "is : ", outputP)
''' Last used to enter values to use if Perceptron works (testing)
    outputP = 1/(lorses_key_output) #Standard formula'''
```

FIGURE 3.3 Example of code for a neural net written in Python (program courtesy of Arthur Arnx, 2019, Jan. 13)

ted by the user. This mimics how people learn a la behavioral theory, where prompting someone verbally, gesturally, visually, physically, proximally, and/or inadvertently increases the chance that they will act out a behavior of interest (e.g., touching their nose).

ChatGPT has been made to respond to sensible strings: its linguistic scope includes ordinary and higher-level programming languages, as well as mathematical formulations. ChatGPT can reportedly also somewhat write dialogue for comics, "tell a joke at least passably well, write an advertising slogan, make stock picks, and imagine a conversation between Xi Jinping and Vladimir Putin" (Strain, 2023).[7] What would this book be like if ChatGPT wrote it?

brain-imaging would augment our understanding of brain (neural) networks supporting complex behavior.

[7] Purportedly, some have even used it to write their wedding vows!

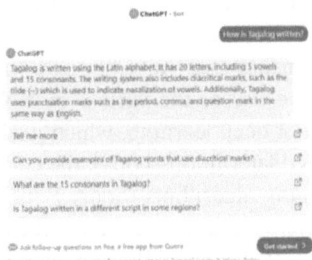

FIGURE 3.4 An example of what ChatGPT can do as a resource for discursive learning

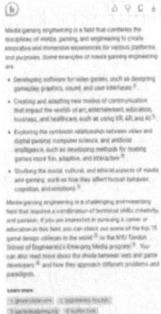

FIGURE 3.5 Bing result for the search query "media gaming engineering" via GPT-4

3.3 EEG AND DEEP CONVOLUTIONAL NNS (DCNNS)

Of especial interest are studies that combine neurotechnologies. Wang and Wang's (2022) work is exemplary of such an approach, as it used both EEG and neural nets (NN). Even more specifically, this study was interested in convolutional NNs (CNNs). Yann LeCun's first CNN was made in 1989 to be applied toward digit recognition. (baeldung noted that CNNs and recurrent NNs, i.e., CNNs and RNNs, are examples of DNNs.) The power of such CNNs lies in their ability to process three types of data: speech, audio, and imagery. Generally, Wang and Wang noted that nearly any NN model in their study performed strongly, given well-formed inputs. This highlights the importance of syntax in computational research.

AugCog could benefit by using a better variety of NNs. Though AugCog uses NNs to superb effect, it relies on the tradition of neural network research as its own entity. This means that AugCog uses NNs but may not be fully aware of their broader history or the variety of NNs available for different tasks. If NN is a subset of deep learning, which itself is a subset of machine learning (this is discussed later in this book), the categorization of CNNs and RNNs as machine learning algorithms (Song, Bang, Choi, & Kim, 2023) is valid. CNN use and reference is a current trend in AugCog, as it featured prominently in research like Wang and Wang's. In 2023's augmented cognition conference proceedings (for HCI International), Rajabi et al.'s study is the only one to both discuss convolutional NNs and cite work about them.

It is possible to cover multiple kinds of NNs in a given study. This is especially pertinent if researchers are interested in comparing the performance, accuracy, or speed of different kinds of NNs. Tavanaei, Ghodrati, Kheradpisheh, Masquelier, and Maida (2019) cover ANNs, DNNs, and SNNs. To the extent that future neural net AugCog research incorporates works from outside of the field, it is possible that AC will have future studies done using (if not at least discussing) multiple kinds of NNs.

Each different type of NN represents a new kind of brain, even if only artificial. As is the case in the animal kingdom, mammals with different sizes of brains occupy different niches and positions in their ecosystems. One could also speak of a technological ecosystem of distinct kinds of NNs, each made to occupy its own niche, or carry out its own task in its own kind of way.

Recent AugCog research has used deep (concurrent) NNs, but there is another kind of DNN that does not seem present in HCI AugCog conference proceedings. This is the *delayed* NN. A subset of this is the delayed Hopfield NN: a subset of this is the delayed periodic Hopfield NN. Also existent are the delayed almost periodic NN and delayed periodic Cohen-Grossberg competitive and cooperative NNs. These NNs have been studied in the area of complex systems. Though not mainstream AugCog topics, my team and I have studied game theory—perhaps the best-known area using the concepts of competitive and cooperative regularly (seen in the prisoner's dilemma)—and complexity.

3.4 RELATED TOPICS

I have already mentioned how neural nets can augment cognition in this book, but there may be other ways they can do so. A more subjective way is to inspire models of cognition. If cognition is indeed a mirror of the brain to some significant extent, and neurons are the basic unit of the nervous system (including the brain), what is the basic unit of the mind? An answer to this using my model of mind is some average of affect, cognition, motivation, and behavior that underlies all of these as more fundamental. For readers who understand statistical procedures used in social science, factor analysis comes to mind as an analogous way to arrive at such a "mental unit" (if one were to perform factor analysis on affect, cognition, motivation, and behavior in some tractable way). Basic units are important for any reductionist science such as particle physics.

Other names for the neural network are *neuronal network* (see Gerstner, Kistler, Naud, & Paninski, 2014) and *cortical network* (as in Fuster, 1999). If memory and perception can be treated as kinds of networks, they are more easily related to NNs. Fuster seems unconcerned about distinguishing between such cognitive and neural networks (he defined memory as "a cortical network" (p. 96)). But even if one is concerned about this, at least the way is paved to discuss neurocognitive networks (perhaps *NCNs*). The more studies and applications of cognition and neurology share language, the more this might be reflected in their realities.

Neural nets (along with genetic algorithms) often have complexity $O(!N)$. This makes them great at passing elements through them but suboptimal at carrying out operations, all in terms of time complexity. Heaton (2013) described ANNs as attempts to model algorithms based on the BNN. For him, neural nets are a "small part" of AI research (p. 6). He describes the ANN (albeit, referring to them simply as a neural network) as aligning with multiple AI algorithms. Such neural nets, Heaton writes, are unlike the human BNN since they are not general computers. Rather, they perform narrow tasks, making them in this sense a form of narrow AI.

Neurons are nerve cells. If NNs are based on networks of nerve cells, they could reasonably be expected to perform the same functions as such cells do. Given nerve cells' primary role in generating sensation, could NNs someday do this? What would it be like for an ANN (for example) to process sensation?[8]

Writing further on the function of ANNs, Heaton noted that "training is the weight matrix". When training takes place is dependent on the algorithm being used. Apparently, "online training support" is frequent with respect to NNs. Heaton noted another kind of NN, i.e., the *deep belief* NN. Highly complex machine learning (ML) algorithms that learn and recognize characters are often NNs. "Bias nodes" are also frequent in NNs (p. 119). Finally, for Heaton's treatment, backpropagation-training algorithms are related to NNs.

---

[8] It has been noted that cognitive scientists "have long treated neural network models of language processing with skepticism" (Schrimpf, Blank, Tuckute, Kauf, Hosseini, Kanwisher, Tenenbaum, & Fedorenko, 2021).

Varela et al. (2000) also discussed NNs, albeit briefly. They wrote that rules like Hebb's rule (which concerns weighting modeled neurons) offer an NN not only including "emergent configurations"—they also give it the ability to create novel configurations that are informed by experience.

RNNs have been given a critical treatment by Kim, Luo, Pillow, and Brody (2021). They cite research proving that such NNs "can be trained to approximate nonlinear latent dynamics" in active brain regions like the "motor and premotor cortices" (p. 5551). They go on to note that RNNs are usually "high-dimensional" and deterministic. "Deterministic latent dynamics" (p. 5552) may work fine for an activity like cycling or any other autonomous motor activity. Also, a formula for "modeling neural population activity in terms of the generic dynamical system" (p. 5551) was approximated via latent "factor analysis via dynamical systems" (LFADS) and an RNN in terms of "the generator":

$$h_0 \sim N\left(\mu_{h_0}, Q_{h_0}\right)$$

$$h_{k+1} = GRU\left(h_k, u_k\right)$$

$$\text{"}z_k = Wh_k + b\text{"} \left(\text{p. } 5552\right)$$

(3.4.1)

LeCun posed three challenges for artificial intelligence and machine learning. His second challenge was for such applications to learn how to reason. He likens such reasoning to economist and influential cognitive theorist Daniel Kahneman's System 2 mode of cognition. In LeCun's vision, AI and ML would go beyond the limits of System 1—fast, intuitive, "feed-forward" (Berkeley EECS, 11:24)—subconscious computing, and learning would be compatible with reason.

Kim et al. note that "LFADS...has a black box recognition model called the encoder that serves to amortize inference, where it gets a sequence of spike counts $x_{1:K}$ as an input and output parameters μ_{h0} and Q_{h0}". u is replaceable with "outputs from an extra RNN called the controller".

The popular song-identifying mobile app Shazam uses a DNN (Tao & Getachew, 2020). This means neural nets' functions go past photograph identification and include auditory identification. However, the principle underlying the former is the same as with the latter. DNNs predict a whole result—e.g., which person is in a photo, or which song is playing audibly in a space—by summing up parts of a sample. In Shazam's case, its predictions are matched with an existing database of songs before being shown to its user. Shazam is a good example of augmented cognition in an artistic domain (music).

The online book *Neuronal Dynamics: From Single Neurons to Networks and Models of Cognition* is both an accessible and a well-suited resource for anyone seeking to understand how neural networks relate to cognition. The connection between these is, understandably, not entirely intuitive (especially for those who view the brain and mind as distinct entities). This book contains back-to-back chapters on neural networks and cognitive dynamics: one could imagine a follow-up chapter on augmenting neural network-related cognition.

Another good resource for further learning on and interacting with neural networks is BrainFacts.org. On this site, one can interact with a virtual model of the human brain and select different regions to learn more about them. The site also has a comprehensive search feature: one could find numerous articles on NNs through it.

Outside of AugCog, Beckmann, Köstner, and Hipólito (2023) proposed a "computational phenomenology" and applied it to deep learning. They proposed this as an alternative to traditional deep learning that relied on symbolic representationalism, the position that cognition primarily consists of mental maps or representations (i.e., how things appear visually in one's mind) and formal operations on humanmade symbols. These authors shift the focus of perception onto experience away from mental maps. Though this theoretical reorientation is arguably only relevant for those seeking to create conscious artificial intelligence, it shows the breadth of thought displayed by researchers in deep learning. Also outside of AugCog, Saminu, Xu, Zhang, Abd El Kader, Aliyu, Jabire, Ahmed, and Adamu (2022) considered the artificial neural network to be a "conventional machine learning" methodology (p. 20). However, NN is not to be confused in this book with the "nearest neighbor" machine learning method.

Modesti, Rapisarda, Capriotti, and Del Casale (2022) conducted a meta-analysis of neuroimaging research (using fMRI and PET-scanning) on

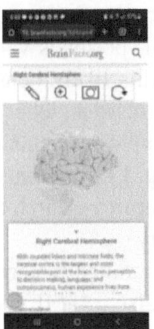

FIGURE 3.6 Image and description of the right cerebral hemisphere (Society for Neuroscience)

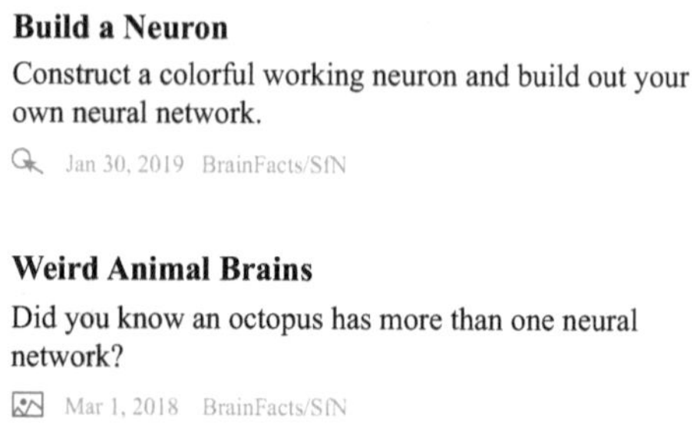

Build a Neuron

Construct a colorful working neuron and build out your own neural network.

Jan 30, 2019   BrainFacts/SfN

Weird Animal Brains

Did you know an octopus has more than one neural network?

Mar 1, 2018   BrainFacts/SfN

FIGURE 3.7 Two search results on BrainFacts.org (Society for Neuroscience resources) for "Neural network". Articles found this way include interesting topics relevant to neural networks, including an educational source on building one's own NN and information about the biological neural networks of non-human animals such as the octopus. (The octopus is an animal often considered surprisingly intelligent—especially for a non-human.) Society for Neuroscience resources like these augment user cognition about neural networks in ways not accessible via the field of augmented cognition.

dissociative dysfunction. In addition to implicating the prefrontal cortex in such disorders, they concluded that alterations of "the functional neural network of the caudate [nucleus]" are relevant to changes in identity and preservation of "altered mental status" in dissociative identity disorder (DID). Identity is not a topic studied in AugCog, but a changed mentality of a more positive sort is what AugCog achieves. AugCog could leverage neurotechnology to understand positive transformations in identity.

Could a myriad of simplistic perceptrons be connected in a vaster neural network resembling more closely a real brain? Should such be done, if so? Undoubtedly, the ability to simulate phenomena like drug absorption via the brain in an artificial one could be good for neuropsychiatry, as anything that reduces organism-testing yet still portends value for science and application is preferable. Still, even given this likely benefit, the brain consists of neurotransmitters, synapses, layers, regions, lobes, and billions of neurons. Such complexity would demand a complex systems approach, which is not usually carried out explicitly in AugCog. I have pioneered some such work (Sood et al., 2019), but it would be up to future research to render such complexity work relevant to the brain in its entirety.

In summary, NNs are programmed to carry out tasks. They may be considered a subset of AI, but they also use ML methods. They use ML methods to the extent that they are trained with data to carry out tasks. Tasks NNs carry out include visual and auditory identification: they may be hooked up with an external database (e.g., a song library) to do the latter.

REFERENCES

baeldung. Multi-Layer Perceptron vs. Deep Neural Network. *Baeldung*, 2023, June 13. https://www.baeldung.com/cs/mlp-vs-dnn

Beckmann, P., Köstner, G., Hipólito, I., An alternative to cognitivism: computational phenomenology for deep learning, *Minds and Machines*, 2023. https://doi.org/10.1007/s11023-023-09638-w

Berkeley EECS, "Yann LeCun: From Machine Learning to Autonomous Intelligence", *YouTube*, September 27, 2022. https://www.youtube.com/watch?v=VRzvpV9DZ8Y&t=555s

Boyd, J. P., Batchelder, W. H., Network Analysis. In: Batchelder, W. H., Colonius, H., Dzhafarov, E. N., Myung, J. (eds) *New Handbook of Mathematical Psychology*, Cambridge University Press, Cambridge, 2016, 194–273.

Chowdhury, J. R., Caragea, C., Beam Tree Recursive Cells, *arXiv*, 2023a. https://arxiv.org/abs/2305.19999

Chowdhury, J. R., Caragea, C., Efficient Beam Tree Recursion, *arXiv*, 2023b. https://arxiv.org/abs/2307.10779v2

Corella, F., Lewison, K., Biometrics. In: Moallem, A. (ed) *Human-Computer Interaction and Cybersecurity Handbook*, CRC Press, Boca Raton, 2019, 458.

Cretu, C., How Does ChatGPT Actually Work? An ML Engineer Explains. Scalable Path, 2023. Retrieved from https://www.scalablepath.com/data-science/chatgpt-architecture-explained

DeepLearning.AI, "Unsupervised Learning, Recommenders, Reinforcement Learning", *Coursera*, 2024. https://www.coursera.org/learn/unsupervised-learning-recommenders-reinforcement-learning

Deng, X., Xu, J.-X., Wang, J., Wang, G., Chen, Q., Biological modeling the undulatory locomotion of *C. elegans* using dynamic neural network approach, *Neurocomputing*, 186, 207–217, 2016.

Fuster, J. M., *Memory in the Cerebral Cortex: An Empirical Approach to Neural Networks in the Human and Nonhuman Primate*, Bradford, Massachusetts, 1999.

Gerstner, W., Kistler, W. M., Naud, R., Paninski, L., *Neuronal Dynamics: From Single Neurons to Networks and Models of Cognition*, Cambridge University Press, 2014. https://neuronaldynamics.epfl.ch/online/index.html

Glenn, J. C., Florescu, E., and The Millennium Project Team, State of the Future V. 19.0, The Millennium Project, 2019.

Heaton, J., *Artificial Intelligence for Humans, Volume 1: Fundamental Algorithms*, Heaton Research, 2013.

Kim, T. D., Luo, T. Z., Pillow, J. W., Brody, C. D., Inferring latent dynamics underlying neural population activity via neural differential equations, *Proceedings of the 38th International Conference on Machine Learning*, 139, 5551–5561, 2021.

Kundariya, H., Top 7 NodeJS libraries and tools for machine learning, datasciencedojo, Top 7 NodeJS libraries and tools for machine learning | Data Science Dojo, 2022. Retrieved from https://datasciencedojo.com/blog/nodejs-libraries-machine-learning/

Minsky, M., "Neural Nets and Theories of Memory", AI Memos (1959–2004), 1963. https://dspace.mit.edu/handle/1721.1/6103

Modesti, M. N., Rapisarda, L., Capriotti, G., Del Casale, A., Functional neuroimaging in dissociative disorders: a systematic review, *Journal of Personalized Medicine*, 12(9), 2022. https://doi.org/10.3390/jpm12091405

Neuroelectrics, "Rosenblatt Perceptron, the origin of Artificial Neural Networks", *Neuroelectrics*, 2023. https://www.neuroelectrics.com/blog/2023/10/03/rosenblatt-perceptron-the-origin-of-artificial-neural-networks/

Poirazi, P., Brannon, T., Mel, B. W., Pyramidal neuron as two-layer neural network. *Neuron*, 37(6), 989–99, 2023. doi: 10.1016/s0896-6273(03)00149-1.

Saminu, S., Xu, G., Zhang, S., Kader, I. A. E., Aliyu, H. A., Jabire, A. H., Ahmed, Y. K., Adamu, M. J. Applications of artificial intelligence in automatic detection of epileptic seizures using EEG signals: a review, *Artificial Intelligence and Applications*, 1(1), 11–25, 2022. https://doi.org/10.47852/bonviewAIA2202297

Schrimpf, M., Blank, I. A., Tuckute, G., Kauf, C., Hosseini, E. A., Kanwisher, N., Tenenbaum, J. B., Fedorenko, E., The neural architecture of language: integrative modeling converges on predictive processing, *PNAS*, 118(45), 2021. https://doi.org/10.1073/pnas.2105646118

Sood, S., The Psychoinformatic Complexity of Humanness and Person-Situation Interaction. In: Arai, K., Bhatia, R. (eds) *Advances in Information and Communication. Lecture Notes in Networks and Systems*, vol. 69, Springer, Cham, 2019. [Also presented at the Future of Information and Communication Conference 2019] https://doi.org/10.1007/978-3-030-12388-8_35

Song, J., Bang, S., Choi, N., Kim, H. N., Brain organoid-on-a-chip: A next-generation human brain avatar for recapitulating human brain physiology and pathology, *Biomicrofluidics*, 16, 2023. https://doi.org/10.1063/5.0121476, 2023.

Strain, M. R., "Kill the Chatbots?" American Enterprise Institute, 2023. Retrieved from https://www.aei.org/op-eds/kill-the-chatbots/

Tao, S., Getachew, Y., High Fidelity Song Identification via Audio Decomposition and Fingerprint Reconstruction by CNN and LSTM Networks, 2020. 38911459.pdf (stanford.edu)

Tavanaei, A., Ghodrati, M., Kheradpisheh, S. R., Masquelier, T., Maida, A., Deep Learning in Spiking Neural Networks, *arXiv*, 2019. https://arxiv.org/pdf/1804.08150.pdf

Varela, F., Thompson, E., Rosch, E., *The Embodied Mind*, The MIT Press, Cambridge, 2000.

Wang, X., Wang, Z., CNN with Self-attention in EEG Classification. In: Kurosu, M., et al. (eds.), *HCI International 2022 - Late Breaking Papers. Multimodality in Advanced Interaction Environments. HCII 2022*. Lecture Notes in Computer Science, vol 13519. Springer, Cham, 2022. https://doi.org/10.1007/978-3-031-17618-0_36

Brain–Computer Interface (BCI)

4

4.1 A WHOLE-BODY APPROACH TO AUGMENTED COGNITION

People have minds. Minds cognize. This is the basis of augmented cognition as a human area and discipline. Sometimes, we mentally will ourselves to do something or our bodies to move—this is the basis of "volitional" BCI.

Thoughts guide us. When we are uncertain of which course of action to take, or of what to say, we pause and think if we desire a strong outcome. With BCI, the situation becomes that of enabling thought to be read and letting the "reader" control one's actions. If a person is unable to move any of their limbs by himself or herself, a thought-reading device can understand their intent, and an intent-to-behavior converter can manifest their thought.

Perhaps the most tantalizing prospect for BCI is the possibility of making everyday tasks easier, including for the unimpaired. With BCI, a human brain interacts via an interface with a computer, with the goal of augmenting user cognition by amplifying their functional (e.g., sensory) capability. Of the three neurotechnologies in this book, BCI is the most directly and holistically cognition-augmenting for its user. In the latest conference proceedings for augmented cognition, a section devoted to chapters dealing with BCI is titled "Brain–Computer Interfaces and Neurotechnology" (Schmorrow and Fidopiastis, 2023), reflecting the thematic appropriateness of the present focus on BCI as neurotechnology in AugCog. (The other three sections in this volume are "Neuroergonomics, Physiological Measurement, and Human Performance", "Augmented Cognition: Evolving Theory and Practice", "Augmented and Virtual Reality for Augmented Cognition", and "Understanding Human Cognition and Performance in IT Security".)

DOI: 10.1201/9781032692982-5

BCI operates by translating a person's intention (measured via their conscious cognition) into "device commands", which are then translated into extension (behavior) (Shih, Krusienski, & Wolpaw, 2012). According to Shih et al., BCI consists of four procedural parts:

1. Signal acquisition, in which brain activity is recorded
2. Feature extraction, in which user intention is learned of
3. Translation, where a person's intention is turned into a command
4. Output, or what behavior the user intended to act out

Feature extraction (2.) is featured in other AugCog research as well, specifically in detecting bots on Twitter (Neumann et al., 2017). Similarity can be seen between the above process and NNs: both consist of input, processing, and output.

The Belgian Jacques Vidal is credited as having invented the phrase "brain–computer interface". The brain–computer interface (BCI) is, alongside neural chips, the most hyped neurotechnology. Elon Musk's company Neuralink recently was approved by the United States Food and Drug Administration to move forward with human clinical trials (Tech Wizard, 2023, June 15). This may signal a renaissance for BCI in a manner reminiscent of how Myspace heralded the social media revolution (in which case, the ramifications for neurotechnological AugCog would likely be significant). The key difference for BCI is its potential to measure bodily (rather than merely brain) nerve activity. The reader interested in BCI used as a prosthetic arm is referred to Dr. Robert Mann's pioneering work on an "elbow that joined an electromechanical device with remnant muscle tissue" (Elemental MIT, 2016, January 27).

David Eagleman's BCI VEST is a famous (albeit, in my experience some years ago, teasingly inaccessible) example. One hypothetically wears such a vest as one normally would, and the upper torso's nerve activity is measured. According to Eagleman, his graduate student, Scott Novich, and he built their device to "provide sensory substitution for the deaf" (Eagleman, 2015, p. 189). It is worth noting that the description of the VEST in Eagleman's book appears under a section titled, "Sensory Augmentation". If operational, VEST would go "beyond [sensory] substitution" and "extend our sensory inventory…. adding new senses to the human repertoire to *augment* [emphasis added] our experience of the world" (p. 188). Eagleman goes as far as to pose streaming content from the Internet directly into consciousness, without the need for computer screens.

Why (else) has BCI been so hyped up? One reason for BCI's hype is that popular presentations of it—as well as, perhaps, the ideal of it—convey that it makes psychokinesis possible. Psychokinesis, a parapsychological phenomenon, involves the unmediated movement of physical objects with the mind.

There is a sense in which BCI is psychokinetic, but the manner in which it reads neural signals and translates a person's volition into bodily motion is complex. The aim of AlterEgo, a BCI that aims to "augment human cognition and abilities", is to translate "subtle internal movements" (intra-vocalization) into output such as the time of day and grocery store item prices (MIT Media Lab, 2018 April 4). At this point, the most level-headed view of such BCI is that the body mediates the relation between mind and world.[1]

In the most recent conference proceedings for augmented cognition at the 2023 Human–Computer Interaction International Conference (HCIIC), "BCI" had 20 total uses (Schmorrow and Fidopiastis, 2023). In the section "Brain–Computer Interfaces and Neurotechnology", there are two out of four titles mentioning brain–computer interface(s). Another study in this section includes *brain–computer interface* as a keyword. Davelaar's (2023) study is titled, "Discovering Mental Strategies for Voluntary Control Over Brain-Computer Interfaces"; Rajabi et al.'s study is "Mental Face Image Retrieval Based on a Closed-Loop Brain–Computer Interface". Davelaar also wrote on neurofeedback learning, a topic intuitively grasped as suited for neurotechnological AugCog. BCI in 2023 was uniquely prominent in neurotechnological AugCog, as neither EEG nor NN enjoyed mention in a section's title.

The above two works represent distinct applications of BCI, and the latter deals with a specialized kind of BCI known as closed-loop. Closed-loop BCI records data from the brain and no other part of the nervous system. Such BCI uses neurofeedback to analyze this data and augment the brain's activity (Belkacem et al., 2023). In Rajabi et al.'s case, the application was enhancing the recall of faces. Another study in the volume just discussed noted that "advancement in computational technology … has led to…growth of research in the field of brain-computer interfaces (BCI)" (Aksiotis, Tumyalis, & Ossadtchi, 2023, p. 3). Aksiotis et al. further wrote that one "of the most promising areas of BCI is the use of adaptive algorithms that are based on brain signals".

What all of the studies in the just-discussed BCI section have in common is not necessarily in their titles. Only two have "brain-computer interface(s)" in them. Each is distinct, with no shared authors among them. Yet since they appear in the same section, what more could be said beyond that they each have to do with BCI (and neurotechnology)? Aksiotis et al. include EEG, as do Teixeira and Gomes (2023) (discussed more, shortly) and Rajabi et al. Rajabi et al. cite work using a CNN and include an image of its architecture. Davelaar discussed EEG and neural networks ("strategy-supporting" ones (Schmorrow & Fidopiastis, 2023, p. 20)). It is important to pause on these four studies since

---

[1] This is in fact what the phenomenological philosopher Maurice Merleau-Ponty believed: that the body is the nexus between mind and the external world.

they represent the closest sample for determining current trends in AugCog. In this way, these are the four most relevant studies covered in this book.

The more ambitious wing of BCI is the proposed mechanism of *sensory substitution*. By this, it is imagined that someone who lacks one sense—say, touch—can compensate for it by use of one (or more) other sense(s). This is a niche application that may or may not be relevant for all people. If one sense can augment another, this is a valid area for augmented cognition. Sensory substitution is somewhat tricky. How could touch (one sense) "pitch in" for smell? In reality, all five of our senses are connected, so sensory substitution could exploit overlap between senses so that capacity in one could cover deficiency in another. One important limit of sensory substitution is that it (at least, nominally) could not account for perception. Perception is theorized to be a superset including sensation, where perception adds the mental interpretation of what one senses. Only if sensory substitution succeeds could we turn to the possibility of a higher, perceptive substitution: but it is unclear what this would consist of, even hypothetically. The clinical context is less about augmenting cognition, and more about using brain technology to experiment with the body to fill a desire or need for restored sensory completion. This is likely too new an avenue to draw conclusions about, but researchers could look to the already-established phenomenon of neuroplasticity.

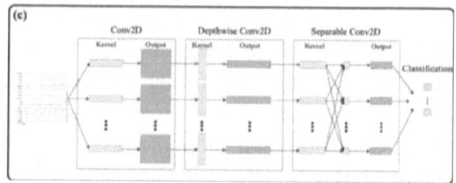

FIGURE 4.1   Structure of the EEGNet CNN (Schmorrow & Fidopiastis, 2023, p. 32)

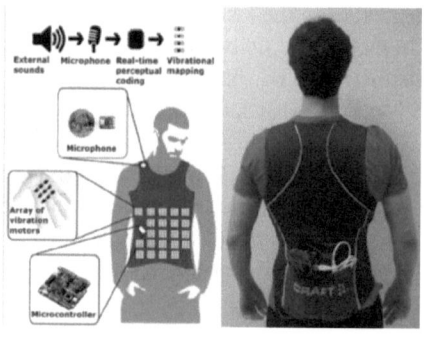

FIGURE 4.2   David Eagleman's BCI VEST (Eagleman, 2023)

Sensory substitution as just described would need to exploit neuroplasticity in order to work. Specifically, brain regions associated with sensation (such as the occipital lobe) could help patch up a deficient neighboring region (like the temporal lobe, involved in processing auditory information). The informatic paradigm enables a more intuitive relation between data technologies and the brain. If the brain is at least partially an information processor, where sensory stimuli are data that are transformed into meaningful information via perception, said information could be passed through informatic neurotechnologies to obtain knowledge.

4.2 AUGMENTED EMBODIED COGNITION

I noted elsewhere that knowledge can be thought of as being encoded in a neurocognitive system. What I had in mind here was, roughly, the brain; or, if the reader here prefers, the "brain-mind". (This hyphenated term might be necessary to denote the philosophical position of monism as opposed to dualism, the latter of which treats the brain and mind as ontologically distinct.) Windhorse and Almadbooh (2022) studied motor-imagery (MI) BCI. MI-BCI is a system where individuals cognize motor activity while not carrying it out. This seems exemplary of BCI applied in the clinical, borderline-paralysis treatment context. Dehaene (2014) noted that strides have been made "decoding auditory attention" (p. 216), as well as in MI. Dehaene additionally noted that BCIs—especially those that use implants—could renew long-distance BNN (interneural) communication.

The relatively recent trend of embodied cognition could be more relevant in AugCog than it has so far been. Work has been done on enactive cognition in relation to virtual reality (VR) (Hovhannisyan et al., 2019), but embodied and enactive cognition could be related explicitly to the neurotechnologies of interest here. Bruineberg and Rietveld (2014) were interested in "how skilled agents interact with their environment and can tend toward *improvement of their situation* [emphasis added]". This evokes AugCog's performance-enhancing imperative somewhat, albeit focusing on the situation rather than behavior. This could represent a new opening in AugCog if Bruineberg and Rietveld's framing of neurodynamics being embedded in "the broader brain-body-environment system" is adopted. Though these authors merely cite works involving EEG and NN, future AugCog research could draw from (or even couch their studies in) the ontology laid out by such progressive cognitive theorists.

BCI has the potential to augment embodied cognition on a general scale. This is because the technology is available in open-source formats. The two most prominent examples of such are BCI2000 and OpenBCI. BCI2000 is a "general-purpose software system for...[BCI] research" (BCI2000 Wiki, 6 February 2024) that is freely usable for non-fiscal purposes. It includes "software tools that can acquire and process data, present stimuli and feedback, and manage interaction with" external devices (e.g., a prosthetic arm). It is "real-time" and can "synchronize EEG and other signals with a wide variety of biosignals and input devices such as...eye-trackers". Eye-tracking is especially relevant for AugCog given its frequent use in cognitive research.

Free use of OpenBCI requires an EEG device independently bought or made. A foundational article regarding OpenBCI is that of Cardona-Álvarez, Álvarez-Meza, Cárdenas-Peña, Castaño-Duque, and Castellanos-Dominguez (2023). These authors describe OpenBCI as offering "unparalleled freedom and flexibility" with its publicly available "hardware and firmware" that are cheaply implementable. Though neither BCI2000 nor OpenBCI was mentioned in 2023's HCI Augmented Cognition proceedings, there is no reason they cannot be used to augment cognition for researchers, appliers, and users of the technology.

4.3 FROM BRAIN TO BODY AND MIND

BCI has the potential to offer an answer to one of the oldest, most elusive questions known to man. What is the relation between thought and reality? Can BCI really give an objective, external reality to our internal and subjective thoughts? And can it really do this by converting a thought into behavior? If BCI can transform thought into behavior, the dualism positing mind and body as separate breaks down. BCI could open the door to a revolution in Western science's epistemology (how we understand or know of things) and ontology (what is).

At this point, the reader may also wonder: can findings about the brain (and body) be mapped onto our understanding of the mind *meaningfully*? This depends on philosophy. If the brain and mind refer to the same entity, findings on the former and latter are one and the same. But to any extent that they turn out to be different, ontologically—meaning that their difference is more than just the words we use to talk about them—we would need distinct sciences for each. Fortunately, neuroscience and cognitive science exist: cognitive neuroscience is a burgeoning area. Technologically, these considerations may not matter much, but they are interesting in their own rights.[2]

---

[2] Robotics is a related area that leverages the brain–computer interface, for instance, using Python (Madden and Dacombe, 2023). Robotic applications of BCI are somewhat harder to

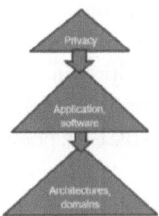

FIGURE 4.3 Index terms represent relevant topics for BCI cybersecurity. Will more intrusive BCI applications leave wearers' minds vulnerable?

Bernal, Celdrán, Pérez, Barros, and Balasubramaniam (2019) point to the importance of using BCI safely. It does not seem well-addressed in AugCog BCI research prior to this book that BCI, and neurotechnology in general, carries a potential cybersecurity risk. This applies if BCIs operate in tandem with the cloud (as Bernal et al. note that BCI is poised to open the door to). If BCI does tap into its user's thoughts, it is obvious that one would not want such a device to be hacked maliciously. BCI might not be as at risk for being hacked given its benevolent application, discussed further in the next section. But use of any technology, especially medically, must lay out any level of risk that might exist for its user.

4.4 MEDICAL HOPE (AND PROMISE)

Guger Technologies—the Austrian company discussed in relation to the EEG typewriter, previously—creates strong BCIs for both invasive and noninvasive research and clinical applications. BCI has shown promise for paralyzed patients by allowing them to think (i.e., intend) to move and do so at small scales. Dehaene (2014) wrote that BCIs could renew some kind of communication for conscious, yet totally paralyzed (locked-in) patients. This is, to my mind, the most interesting or at least useful application of any discussed in this book. Not only does it shed light on the mind–body problem discussed so commonly in philosophy of mind—it offers a pathway to clinical research and betterment hitherto unprecedented. BCI unseats the dogmatic position of

---

classify as augmented cognition, since studies in the area (under human–computer interaction) do not often include these. Still, they are interesting applications in their own right and are related to clinical applications that attempt to read a user's mind, a kind of application more germane to AugCog.

epiphenomenalism, a term stressing one-way causation of the brain (as cause) on the mind (as effect). Indeed, it switches this common formulation, highlighting the important role of cognition in embodiment. There is a case to be made that thought can alter the brain, which becomes relevant for theory in augmented cognition. If—via MI-BCI—a person can move simply by thinking (that is augmented in the process), what else can they do? Imagination truly becomes the limit!

Teixeira and Gomes (2023) noted that: "Brain-Computer Interaction (BCI) technology can be used in several areas and has recently gained increased interest with diverse applications in the area of Human Computer Interaction (HCI)" (p. 47). They also mentioned its role in neuromarketing, specifically in how the brain responds to "marketing stimuli" (p. 47). BCI research in AC usually seems to use EEG in tandem. BCIs themselves are used mainly to augment user performance based on their cognition. BCIs record and interpret a wearer's neural signals to infer what they are thinking. In this way, BCI is the most ambitious of these three neurotechnologies and attempts to be the most directly cognitive.

At the conclusion of Section 4.3, a potential ethical concern was raised for BCI in relation to cybersecurity. This concern could be grouped under the umbrella of neuroethics, which is concerned with the correct implementation and a conscious advancement of neural science. (Here, I treat EEG, NN, and BCI as being adjacent to or possibly part of neural science.) Ethics is usually about the means and ends of action. Thus, neuroethics is about the means and ends of neural science. How could we augment the ethical research, development, and application of EEG, NN, and BCI? More broadly, to what extent is augmenting cognition via neurotechnology ethical? Pursuing medical ends already seems like an ethical prerogative. Neuroethics is relevant to this book's discussed trends since both are relevant to "near-term innovations in brain science" ("Neuroethics").

In general, Greene (2021 May 21) noted that "it seems like BCIs are a shoe-in to become the next big thing in tech. It's even arguable they could become mainstream before driverless cars do…. On the other hand, it could take decades". This leaves the status of future BCI trends in AugCog indeterminate, given that the state of BCI applications in more mainstream settings is so. AugCog BCI research could become more paradigmatic, ideally in a way encompassing different kinds used. Xie, Xu, Luo, Li, Zhang, Han, and Yan (2017) discussed the "spatial selective attention-based brain-computer interface (BCI) paradigm" that is "steady-state visual evoked potential (SSVEP) BCI". AugCog BCI research need not be as specific as this in outlining its own paradigm, but a unifying theoretic structure should be useful to have.

REFERENCES

Aksiotis, V., Tumyalis, A., Ossadtchi, A. Brain State-Triggered Stimulus Delivery Helps to Optimize Reaction Time. In: Schmorrow, D. D., Fidopiastis, C. M. (eds) *Augmented Cognition. Lecture Notes in Computer Science*, vol. 14019, Springer, Cham, 2023. https://doi.org/10.1007/978-3-031-35017-7_1

BCI2000 Wiki, *BCI2000*, 2024, 6 February. https://www.bci2000.org/mediawiki/index.php/Main_Page

Belkacem, A. N., Jamil, N., Khalid, S., Alnajjar, F., On closed-loop brain stimulation systems for improving the quality of life of patients with neurological disorders. *Frontiers in Human Neuroscience*, 17, 2023. https://doi.org/10.3389/fnhum.2023.1085173

Bernal, S. L., Celdrán, A. H., Pérez, G., Barros, M. T., Balasubramaniam, S., Security in brain-computer interfaces: state-of-the-art, opportunities, and future challenges, *ACM Computing Surveys*, 2019. https://doi.org/10.1145/3427376

Bruineberg, J., Rietveld, E., Self-organization, free energy minimization, and optimal grip on a field of affordances, *Frontiers of Human Neuroscience*, 8, 599, 2014.

Cardona-Álvarez, Y. N., Álvarez-Meza, A. M., Cárdenas-Peña, D. A., Castaño-Duque, G. A., Castellanos-Dominguez, G., A novel OpenBCI framework for EEG-based neurophysiological experiments, *Sensors*, 23(7), 3763, 2023. https://doi.org/10.3390/s23073763

Davelaar, E. J., Discovering Mental Strategies for Voluntary Control Over Brain-Computer Interfaces. In: Schmorrow, D. D., Fidopiastis, C. M. (eds) *Augmented Cognition. Lecture Notes in Artificial Intelligence*, vol. 14019, Springer, Cham, 2023, 16–25.

Dehaene, S., *Consciousness and the Brain: Deciphering How the Brain Codes Our Thoughts*, Viking, New York, 2014.

Eagleman, D., *The Brain: The Story of You*, Vintage Books, New York, 2015.

Eagleman, D., Sensory Substitution, eagleman.com. https://eagleman.com/science/sensory-substitution/ (accessed 5 Sept 2023)

Elemental MIT, "Robert W. Mann's 'Boston Arm', *YouTube*, 2016, 27 January. https://www.youtube.com/watch?v=ys9fFJf8pMw

Greene, T., "What would happen if we connected the human brain to a quantum computer?", *TNW*, 2021, 21 May. https://thenextweb.com/news/what-would-happen-interface-human-brain-ai-quantum-computer

Hovhannisyan, G., Henson, A., Sood, S. Enacting Virtual Reality: The Philosophy and Cognitive Science of Optimal Virtual Experience. In: Schmorrow, D., Fidopiastis, C. (eds) *Augmented Cognition. Lecture Notes in Computer Science*, vol. 11580, Springer, Cham, 2019. https://doi.org/10.1007/978-3-030-22419-6_17

Madden, A., Dacombe, S., *The Python Book*, Future Publishing Ltd., UK, 2023.

MIT Media Lab, AlterEgo: Interfacing with devices through silent speech, *YouTube*, 2018 4 April. https://www.youtube.com/watch?v=RuUSc53Xpeg (accessed 16 September 2023).

Neumann, S., *et al.* Content Feature Extraction in the Context of Social Media Behavior. In: Schmorrow, D., Fidopiastis, C. (eds) *Augmented Cognition.*

Neurocognition and Machine Learning. Lecture Notes in Computer Science, vol. 10284, Springer, Cham, 2017. https://doi.org/10.1007/978-3-319-58628-1_42

Neuroethics, *BrainMind*, n.d. https://brainmind.org/neuroethics

Rajabi, N., Chernik, C., Reichlin, A., Taleb, F., Vasco, M., Ghadirzadeh, A., Björkman, M., Kragic, D., Mental Face Image Retrieval Based on a Closed-Loop Brain-Computer Interface. In: Schmorrow, D. D., Fidopiastis, C. M. (eds) *Augmented Cognition. Lecture Notes in Artificial Intelligence*, vol. 14019, Springer, Cham, 2023, 26–45.

Schmorrow, D. D., Fidopiastis, C. M. (eds), *Augmented Cognition*, Springer, Cham, 2023.

Shih, J. J., Krusienski, D. J., Wolpaw, J. R., Brain-computer interfaces in medicine, *Mayo Clinic Proceedings*, 87(3), 268–279, 2012.

Tech Wizard. Elon Musk's Neuralink Brain Chip Firm Wins US FDA Approval for Human Study! *YouTube*. 2023, 15 June. https://www.youtube.com/watch?v=EtRfeAyCV2w&t=4s

Teixeira, A. R., Gomes, A. Analysis of Visual Patterns Through the EEG Signal: Color Study. In: Schmorrow, D. D., Fidopiastis, C. M. (eds) *Augmented Cognition. Lecture Notes in Computer Science*, vol. 14019, Springer, Cham, 2023. https://doi.org/10.1007/978-3-031-35017-7_4

Xie, J., Xu, G., Luo, A., Li, M., Zhang, S., Han, C., Yan, W., The role of visual noise in influencing mental load and fatigue in a steady-state motion visual evoked potential-based brain-computer interface, *Sensors,* 17(8), 2017. https://doi.org/10.3390/s17081873

Windhorse, Y., Almadbooh, N., Optimizing ML Algorithms Under CSP and Riemannian Covariance in MI-BCIs. In Kurosu, M., et al. (eds.), *HCI International 2022 – Late Breaking Papers. Multimodality in Advanced Interaction Environments. HCII 2022.* Lecture Notes in Computer Science, vol. 13519, Springer, Cham, 2022. https://doi.org/10.1007/978-3-031-17618-0_38

Exploring a Combinatory Approach

5

5.1 SYNTHESIS (THEORY AND TECHNOLOGY)

Each of the three neural technologies focused on in this book—EEG, NN, and BCI—offers something unique to augmented cognition. EEG helps us study brain regions; NN lets us simulate real brains via simple software modeling; BCI focuses on the nexus between brain and computer (along with its tremendous clinical potential). It would be worth enumerating each of the possible ways NN, EEG, and BCI are and can be combined. This includes combining NN with EEG, EEG with BCI, BCI with NN, and NN, EEG, and BCI all together.

EEG and BCI have both been studied in a linguistic context by Moreira et al. I was a participant in this study from 2021 to 2022. (This study used transcranial magnetic stimulation—i.e., TMS—as well.) Aksiotis et al.'s study cited in the last chapter used, in addition to BCI, offline EEG (and EMG—electromyographical) processing, representing yet another case of EEG and BCI used in tandem. BCI turns thought into action; EEG scans and makes images of the brain; NN mimics the brain and can act as a mind.[1]

Revlin (2013) described EEG in a manner evocative of BCI. He noted how EEG has been used to facilitate communication between "paralyzed people" (p. 36). Specifically:

---

[1] Though this might seem controversial, the strand of artificial intelligence research and development known as *artificial general intelligence* has such ambitions!

DOI: 10.1201/9781032692982-6

If paralyzed individuals are hooked up to a computer that uses an EEG to record their brain waves...[they] can select letters displayed by an assistant merely by thinking of something that makes them tense. This changes the brain waves, and the computer informs the assistant that the letter has been chosen.

Could the human brain be considered a neurotechnological BNN? If so, studies like those using EEG-BCI systems to study the brain combine all three main neurotechnologies discussed in this book.

Mental stimulation, when technologically applied, augments cognition, affect, motivation, behavior, or any combination of two or more of these. It is appropriate to discuss how powerful the mind is, both at baseline and when cognition is augmented. AugCog can utilize the concept and measurement of the intelligence quotient (IQ), at least indirectly, for inspiration. Since IQ was initially used as a measure of a person's or group's mental age divided by their chronological age, it gives a clue to the AugCog researcher seeking quantification of cognitive augmentation.

I have proposed a humble and down-to-earth way to quantify cognition, simply by counting the number of discrete thoughts that occur. Such quantification can be a function of time, where one may measure how many cognitions occur during a given interval. This could be a basis for quantifying augmentation: if cognition is countable, then it is worth asking to what extent the augmentation of it is (or is not). Cognitive augmentation can enhance cognition's quality or boost its quantity. The important question practically is how (much) cognition needs to be augmented to carry out a task that resists ordinary cognitive effort. Cognitive augmentation can enhance cognition at any level, from dysfunction (which can be conceptualized numerically as negative functioning) to baseline (a neutral cognitive state) up to extraordinary (positive) functioning.

Convolutional neural nets (again, CNN) were studied in tandem with EEG by Wang and Wang (2022). They were also interested in how self-attention worked in relation to BCI. Thus, Wang and Wang's study represents an important one for the study of the three neurotechnologies discussed in this book. Another trend can thus be said to emerge between 2022 and 2023 AugCog research, which is the study of all three neurotechnologies within one study.

At this point, the easy synergy of neurotechnologies should be apparent to the reader. Both EEG and (C)NN research have been related to speech decoding and processing, respectively. A trend in both EEG and NN AugCog research from 2022 to 2023 is classification. Wang and Wang implemented "transformer-based classifiers" in EEG "classification tasks". Rajabi et al. (2023) used an "EEG signal classifier" along with three other components (Schmorrow & Fidopiastis, 2023, p. 31). Provocatively, Dehaene (2014) wrote of neurotechnologies that future ones will alter the landscape of clinical consciousness disease treatment permanently. EEG and BCI can already be used to study BNNs. Could they someday enable the study of ANNs, especially as implemented into robots?

EEG and NN were also studied together in HCI more broadly by Kleybolte and Märtin (2023). They conducted real-time emotional recognition using an EEG and deep NN on Raspberry Pi. Outside of AugCog, Saminu et al. (2023) used deep learning and EEG in a single study to detect epileptic seizures automatically. Even different NNs can be combined, e.g., CNN with RNN.

Dehaene (2014) discussed BCI that is inspired by the more common EEG. All EEG needs is the amplifying of electric signals read from the scalp, which is important given EEG's lower spatial resolution for small electrical impulses. The brain's surrounding "skull and protective tissue" make electrical signals less accessible for EEG (Willingham, 2007, p. 54). EEG as BCI has been described by NeuroTechX with the example of the machine called "P300". P300 affords both words and letters to be spelled using ERPs. (EEG has also been combined with MRI.) Teixeira et al. discussed the neural processing of color as a topic that EEG has been applied toward (to better understand the phenomenon) before noting BCI's use in neuromarketing. Here is just another example of how EEG and BCI can be involved in the same study; Teixeira's study used a BCI-EEG device classified as "acquisition equipment, in the form of a headband, called InteraXon's MUSE 2". They noted MUSE 2's strengths being its practicability and flexibility.

The other most important study for this book is Rajabi et al.'s (2023). Like Davelaar's study, Rajabi et al. discuss all three of the neurotechnologies focused on here. In fact, they mention all three in one sentence, noting that they used "EEGNet…a deep convolutional neural network, as the classifier, as it performs well on many BCI tasks" (Schmorrow & Fidopiastis, 2023, p. 33). Rajabi et al. recruited 17 young adult participants: EEG was recorded "using 32 Ag/AgCl BrainProducts active electrodes" (p. 35). They employed a closed-loop task, proposing a "closed-loop BCI framework" using EEG signaling to ascertain "mental face image" in participants' minds probabilistically (p. 40). The authors also covered previous work that used a deep CNN "to encode EEG signals" (p. 30). Though Rajabi et al. only combine EEG and BCI, it is appropriate to wonder if they could also use NN in future work, finding a novel way to use it with their EEG-based BCI.

Mathematics has also played a role in EEG-BCI research. Congedo, Barachant, and Bhatia (2017) lay out the Riemannian geometry that has increasingly been used in BCI decoding. These authors also have an algorithmic interest in the framework, offering both Python and MATLAB "code libraries for designing BCI decoders". This work is worth studying not only because it is relevant to at least two of this book's three neurotechnologies,

FIGURE 5.1 Using a headband like MUSE, the stars represent channels and the square represents the reference node for electrodes on the scalp (e.g., as in Schmorrow & Fidopiastis, 2023, p. 48). Davelaar's (2023) study was discussed in the Chapter 4. This is one of the most important studies covered in this book. Not only does it appear in the most recent volume of HCI-augmented cognition conference proceedings—it also includes all three of the main neurotechnologies covered here. Thus, it represents the synthesis that the current book section is about. Davelaar noted EEG's being "the most common neural modality" (Schmorrow & Fidopiastis, 2023, p. 16). (He also covers EEG work done by others studying the frontal lobe.) This work includes a figure of an NN architecture (see fig. in Chapter 4) "capable of selecting and deselecting specific representations while being given a general prediction error signal" (p. 20).

but also because it is a cited work in the most recent conference proceedings for augmented cognition. It is a good resource for understanding the relation between geometry and the data science of BCI.

Thus far, it seems that EEG has been used with BCI and with NN. What about BCI and NN? And could all three be used in a single study or as part of a singular application? An EEG-BCI-NN hybrid in AugCog would:

1. Be wearable and usable via software
2. Render brain data and imagery, augment cognition (e.g., perceptual senses), and control its organism.

BCI and EEG are wearable; NN isn't. NN is simultaneously the most abstract neurotech of these three (when artificial) and the closest to the human (where biological). BNNs are the most natural neurotechnology discussed here. A combination of BCI and EEG is possible if one wears an EEG helmet at the same time as the BCI VEST. This along with the brain being a BNN is the easiest combination of the three neurotechnologies discussed. It would be interesting if one could hook up an ANN to a human through some intermediary hardware. If this were possible, one could attempt to program an ANN to

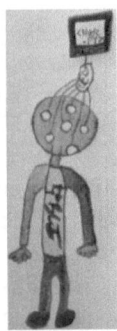

FIGURE 5.2 An imagined rendering of an NN-powered robot wearing an EEG helmet (connected to the "PCtosh", which is running a "Node.py" script) and a sleeveless BCI shirt

mimic a human cognitive module and then use it to cognitively substitute for missing cognition in a human subject or augment one or more existing modules. This would combine ANN with BNN in a way that thus far does not seem to have been done. Another way to combine all three neurotechnologies would be to use an NN model based on an EEG brain scan that informs a worn BCI of the brain state to be augmented.

Researchers could even attempt to brain-image a robust NN using EEG. It is worth mentioning the flexibility of the NN concept here, as it need not have to do with the brain: it could (be a model of) any nerve cell or cluster. Thus, it affords a potentially broad amount of modeling inspiration to the aspiring technologist who wants to understand select functions of the nervous system. EEG rigs could be considered BCIs to the extent that the former's readings are outputted to a computer in real time (as they were when I had my last EEG scan).

5.2 COMPUTATIONAL NEUROSCIENCE'S "AUGMENTED MIND-BRAIN"

For the sake of this book's treatment, AugCog could be regarded as overlapping with and extending beyond computational neuroscience. Here is the most direct way neurotechnologies can augment cognition. By exploiting knowledge of cognitive neuroscience, neurotechnologies can target specific brain regions to augment the kinds of cognition associated with each. EEG data of prefrontal cortex (PFC) activity can be used to augment deliberative thinking, strategizing, and decision-making—executive functioning—enabled by the PFC. The same can be done for the occipital, temporal, and parietal lobes so that visual, auditory, and spatial (respectively) cognitions can be augmented. Neocortical activity can be measured and higher-level, abstract thought and metacognition can be augmented.

Neural nets can augment cognition by enhancing our knowledge of human emotion. For a conference somewhat adjacent to AugCog, Leung and Xu (in press) studied facial recognition in a unique way. They fed photos of human faces to large-language models—ChatGPT-4, in particular—to see how well they could predict a person's emotional state at the time of photography. Their results were not as robust as NN research in AugCog summarized in Chapter 3, with GPT-4's predictive accuracy being between 66-75.3% in emotional recognition tasks (for datasets labeled emotion detection, facial expression, and neutral-human). These are far from the 98–99% accuracy figures displayed by convolutional NNs in Wang and Wang (2022), but they are promising enough that emotional recognition via NNs is done between 16–25.3% better than chance.

The best way for ANNs to advance science is in their native discipline (computational neuroscience). They should be used to validate "neurally testable hypotheses" (Fuster, 1999, p. 83) that are too problematic to test on brains. Fuster wrote that the best "associate models of memory" that resemble most closely how the brain actually works are founded on the notion of representations of neural connections that are flexible (p. 85). In terms of NNs, such modifiability may mean that neural connections update if a stronger association becomes available that makes a memory better retrievable. However, Fuster was careful, having noted also that such memory models do not assume that the brain functions similarly to a digital computer. Neural memory, he wrote, has also not been assumed to be "comparable to computer memory" (p. 86). The brain is enormously capable of performing parallel processing, in contrast to the typical computer.

5.3 AUGMENTED BRAIN, MIND, AND BODY

Brain, mind, and body make up a potential wellness triad. This is not a wellness book, but the general topic of wellness has trended as of late. As populations recover from the recent COVID-19 pandemic, the question of how to thrive becomes clear once more. Mind–body health should undoubtedly be one of augmented cognition's priorities, particularly with respect to BCI. To miss this opportunity to better human lives with technology would be to miss out on augmented cognition's great potential.

Two of the philosophically most interesting aspects of EEG and BCI, respectively, are free will and consciousness. EEG has been used to try to figure out if consciousness comes before or after motor activity. That is to say: are participants' actions free, or are they predetermined? (As usual, Kaku wrote a good summary of such work—carried out famously by Dr. Benjamin Libet.) These sorts of studies are motivated by interesting philosophical considerations, but it is possible they do not invest enough in underlying theory. Consciousness and the body co-occur, rather than one necessarily preceding the other. Furthermore, the research in question seems to operationalize decision-making behaviorally rather than mentally, and certainly not phenomenologically. Studies like these presuppose (and, indeed, conclude) that either bodily activity or conscious volition must come first. I maintain a correlative view of such phenomena as co-arising, as a more realist or Buddhist ontology supposes. This being said, phenomenology has been shown not to be totally intractable in augmented cognition research (see Hancock and Higley, 2014).[2]

Neural chips are commonplace in fiction, and Neuralink could lead the charge in making them a commonplace fact. Direct attachment to the brain is the closest form of neural augmentation possible, short of implanting devices into or on neurons, neurotransmitters, or synapses. Brain technologies, and technologies in general, have associations that are spiritual. Transhumanism is one such ideological movement. Being mindful of thought is a form of metacognition. This means that cognitive mindfulness falls under the purview of AC, and that neurotechnologies could attempt to be applied toward building it. Improved wellness and mental health promoted by spiritual and religious practices of meditation and prayer are another example of technologically mediated or non-mediated augmented cognition. Yet such practices even augment affect and possibly even behavior.

EEG researchers in AugCog could expand work done to understand spiritual practices like meditation (which have proven to augment cognition in their own right). A good video showing what can be done with EEG in this realm shows spiritual philosopher Ken Wilber willfully altering his brain waves (Integral Life, 2006 November 20). Neural nets could simulate a meditative

---

[2] What might turn out to be less tractable for brain technologies are more spiritual and theological domains. There have been naïve attempts to identify the soul in the brain. Correlates of consciousness have been found, but this pertains to consciousness simply as awareness. It can still be treated as metaphysical, but spirit and soul are phenomena that might be more difficult to correlate exactly with brain regions or with the nervous system in general. Nonetheless, spirituality and religion are important for augmented cognition to the extent that faith and conviction (e.g., as functions of human–computer interaction) lead to stronger performance.

brain state, which would be AugCog to the extent that such a state is optimal for one's thinking either during a task or not. It is too early to say how BCI could assist in the user's spiritual practice, but it could help those with bodily disabilities participate in practices like yoga.

REFERENCES

Aksiotis, V., Tumyalis, A., Ossadtchi, A. Brain State-Triggered Stimulus Delivery Helps to Optimize Reaction Time. In: Schmorrow, D. D., Fidopiastis, C. M. (eds) *Augmented Cognition. Lecture Notes in Computer Science*, vol. 14019, Springer, Cham, 2023. https://doi.org/10.1007/978-3-031-35017-7_1

Congedo, M., Barachant, A., Bhatia, R. Riemannian geometry for EEG-based brain-computer interfaces; a primer and a review, *Brain-Computer Interfaces*, 4(3), 155–174, 2017. https://doi.org/10.1080/2326263X.2017.1297192

Davelaar, E. J., Discovering Mental Strategies for Voluntary Control Over Brain-Computer Interfaces. In: Schmorrow, D. D., Fidopiastis, C. M. (eds) *Augmented Cognition. Lecture Notes in Artificial Intelligence*, vol. 14019, Springer, Cham, 2023, 16–25.

Dehaene, S., Consciousness and the Brain: Deciphering How the Brain Codes Our Thoughts, Viking, New York, 2014.

Fuster, J. M., *Memory in the Cerebral Cortex: An Empirical Approach to Neural Networks in the Human and Nonhuman Primate*, Bradford, Massachusetts, 1999.

Hancock, M., Higley, M. Mining and Modeling the Phenomenology of Situational Awareness. In: Schmorrow, D. D., Fidopiastis, C. M. (eds) *Foundations of Augmented Cognition. Advancing Human Performance and Decision-Making through Adaptive Systems. Lecture Notes in Computer Science (LNAI)*, vol. 8534, Springer, Cham, 2014. https://doi.org/10.1007/978-3-319-07527-3_12

Integral Life, Ken Wilber Stops His Brain Waves, *YouTube*, 2006, 20 November. https://youtu.be/LFFMtq5g8N4?si=9O6qqLk7z9MWYtXG

Kleybolte, L. A., Märtin, C. A Novel EEG-Based Real-Time Emotion Recognition Approach Using Deep Neural Networks on Raspberry Pi. In: Kurosu, M., Hashizume, A. (eds) *Human–Computer Interaction. Lecture Notes in Computer Science*, vol. 14012, Springer, Cham, 2023. https://doi.org/10.1007/978-3-031-35599-8_15

Leung, C., Xu, Z., Emotional recognition and classification using large language models, in press.

Rajabi, N., Chernik, C., Reichlin, A., Taleb, F., Vasco, M., Ghadirzadeh, A., Björkman, M., Kragic, D., Mental Face Image Retrieval Based on a Closed-Loop Brain-Computer Interface. In: Schmorrow, D. D., Fidopiastis, C. M. (eds) *Augmented Cognition. Lecture Notes in Artificial Intelligence*, vol. 14019, Springer, Cham, 2023, 26–45.

Revlin, R., *Cognition: Theory and Practice*, Worth Publishers, New York, NY, 2013.

Saminu, S., Xu, G., Zhang, S., Kader, I. A. E., Aliyu, H. A., Jabire, A. H., Ahmed, Y. K., Adamu, M. J., Applications of Artificial Intelligence in Automatic Detection of Epileptic Seizures Using EEG Signals: A Review. *Artificial Intelligence and Applications*, 1(1), 2022, 11–25. https://doi.org/10.47852/bonviewAIA2202297

Schmorrow, D. D., Fidopiastis, C. M. (eds), *Augmented Cognition*, Springer, Cham, 2023.

Teixeira, A. R., Gomes, A. Analysis of Visual Patterns Through the EEG Signal: Color Study. In: Schmorrow, D. D., Fidopiastis, C. M. (eds) *Augmented Cognition*. *Lecture Notes in Computer Science*, vol. 14019, Springer, Cham, 2023. https://doi.org/10.1007/978-3-031-35017-7_4

Wang, X., Wang, Z. CNN with Self-attention in EEG Classification. In: Kurosu, M., et al. (eds) *HCI International 2022 - Late Breaking Papers. Multimodality in Advanced Interaction Environments. Lecture Notes in Computer Science*, vol. 13519. Springer, Cham, 2022. https://doi.org/10.1007/978-3-031-17618-0_36

Willingham, D. T., *Cognition: The thinking Animal,* 3rd edn., Pearson Prentice Hall, New Jersey, 2007.

Conclusion

It has been said that the past shapes the present. The philosophical doctrine of determinism holds that the past, plus the laws of reality, creates what we perceive as real now. Theoretically, this should apply to any trends we observe in the present. It has also been said (sung, in a song) that the past is merely the future, but known. By this, one who knows the past also knows the future (and logically, also the present, future to the past). The present is defined relatively. The present in which these words are being written differs from the future present in which others besides me will be reading this. An interesting relation exists between current and future trends. On the one hand, they are temporally proximal to one another (as opposed to the past and future, which are divided by the present). Current trends that persist will give way to what turn out to be future trends.

Summarizing what was just said: the past and future may be reflections of one another, albeit with differing experiences or epistemologies. We lived through the past, which—if recorded properly—can be remembered; we may foresee the future, but do not know it until it has passed. As this book is about trends, it should be addressed how trends are established. Trends reflect patterns observed over time. Trends may rise, or they may wane as they give way to newly emerging trends. It should be considered how quickly trends change, as well. Trends may last for any period of time: what usually makes them trends is that they become more noticed by a widening set of people. Trends may be conceptual. Storey, Lukyanenko, and Castellanos (2023) discussed trends in relation to emerging technologies. Emerging technologies are an important concept for this book, since BCI is considered such a technology.

The use of neurotechnologies like EEG in AugCog is itself a trend, as studies are published in its conference proceedings annually. Trends can also be more specific: for instance, how many of these studies also incorporate BCI? Are there more or fewer HCI AugCog studies on any neurotechnologies compared with others? It is fairly easy to note that EEG is the most used neurotech in this space, followed by BCI. It is interesting but somewhat understandable that NNs are not as common. EEG is more widespread, accessible, and established than BCI. However, NNs may be *too* accessible: it is likely that

DOI: 10.1201/9781032692982-7

the number of AugCog publications not featuring them does not reflect how interested AugCog researchers are in them. NNs are easiest for any person to begin tinkering with, as one does not need special hardware past a decently functioning PC to do so. There is no need for NN users to buy extra hardware in the forms of helmets, vests, etc. It may be perceived to be not novel or groundbreaking enough to publish one's independent results creating or using an NN.[1] NN's relative lack in AugCog compared with the other two neuro-technologies discussed in this book may change given the popularity of newer applications like ChatGPT.[2]

So, how are trends to be determined? In this case, we are interested in a certain kind of technology (neurotechnology) in a given context (AugCog). Given that brain technologies are relatively new—compared with, say, writing technology—our relevant timeline is relatively short, starting from the inception of any of the three neurotechnologies this book deals with. Current trends are easy to establish: we need only look at the present year for innovations and applications. We may look at neurotechnology trends in AugCog as they pertain to theory, science, engineering, and other applications (viz., medical). Are there any patterns in neurotechnological development such as: *AugCog neurotechnology* X *development has "on" and "off" years*?

We could even consider recreational trends if any of the AugCog community uses these techs as a hobby. For future trends, we could see what is working best now and project that it will continue to succeed. But we could also look back in increments of two, three, five, or ten years to use the past as a guide for what is to come. This raises the interesting consideration of what the relation is between past and present. The farther forward we try to forecast, the more speculative any trends we predict become. There should thus be practical constraints placed so our trend forecasts are meaningful. More basically, we should also resolve questions like: why is a given AugCog neurotechnology succeeding? Why will or won't it, moving forward? Factors like the global market then become important. Market trends, viz. investment in neurotechnology, will influence AugCog neurotech use and development. Neurotechnology in AugCog could be viewed as a function of the market, which itself is tied to a geopolitical state of affairs. What is most within reach for neurotechnology innovation? What is demanded from AugCog? What demands are being placed on AC neurotechnologists? Where is the community's interest trending toward?[3]

To forecast, should we use a general model for predicting technological innovation? If so, we could apply a mathematical rule like Moore's Law (which

---

[1] This is not to say that novelty cannot exist in how one codes or applies an NN.

[2] Technically, GPT stands for *Generative Pre-trained Transformer.*

[3] Sentiment analysis could also help answer this.

concerns computer chip innovation over time). Moore's Law is relevant to the three neurotechnologies of interest here, since they all are at least somewhat computational (using microchips). How far back do we need to go in the past to forecast future trends? If philosophical determinism is true, and the past plus present lead to a fixed future, future trends should be predictable. There seems to be no such law for neurotechnology at present, let alone AugCog neurotechnology in particular. My ambition here has been to start the quest for one (or more), not to fully formulate such a law. AugCog innovation can be spurred by curiosity or accelerated to meet a demand arising from an emergency. In the latter case, AugCog neurotechs could even be authorized for emergency use! What event(s) could warrant such authorization?

My approach to establishing HCI AC trends in the three main neurotechnologies discussed in this book includes perusing recent HCII conference proceedings. Drawing from an available sample, including proceedings from conferences I participated in (as an author, presenter, and/or program board member), I searched for the following terms: *EEG*, *BCI*, and *neural net*. EEG and BCI are used in all proceedings that used electroencephalography and brain–computer interface. Neural net was the term of choice for NNs since the use of *NN* is not pervasive. Searching for "neural net" yielded results including *neural network*. Following is the amount each of EEG, BCI, and neural net was used in a given year's HCI AC conference proceedings: Figure c.1

- 2015—EEG: 535; BCI: 301; neural net: 13
- 2017—*Augmented Cognition: Enhancing Cognition and Behavior in Complex Human Environments*—EEG: 329; BCI: 466; neural net: 8
  - "EEG BCI"
- 2019—*Augmented Cognition*—EEG: 226; BCI: 105; neural net: 17
- 2020—*Augmented Cognition: Theoretical and Technological Approaches*—EEG: 273; BCI: 47; neural net: 61
- 2022—*Augmented Cognition*—EEG: 199; BCI: 35; neural net: 23
- 2023—*Augmented Cognition*—EEG: 171; BCI: 20; neural net: 17

The following questions emerge from the above data:

- Why have EEG and BCI mentions mostly gone down?
- Why has NN mention mostly gone up?
- Why did BCI mentions get so high in 2017, and why did EEG mentions break their decreasing trend from 2019 to 2020?

These questions could be answered by interviewing the researchers who used these AugCog neurotechnologies. The above frequency data mostly clarify

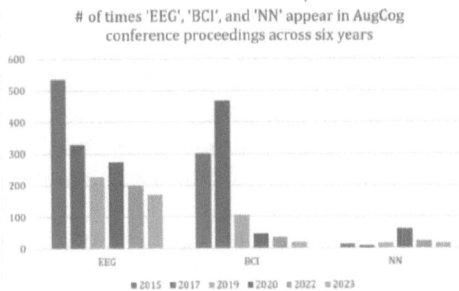

# of times 'EEG', 'BCI', and 'NN' appear in AugCog conference proceedings across six years

FIGURE C.1 Bar chart showing frequency of neurotechnology term usage in AugCog. Years from left to right are in ascending order for each neurotechnology (i.e., 2015 is represented by the farthest-left bar; 2023 is represented by the farthest-right bar).

The years used reflect the proceedings on hand for the author at the time of writing.

what has been said elsewhere in this book. In order from most used to least used in recent AugCog are EEG, BCI, and NN. When I speculated along these lines before, it was before knowing these numbers: to a certain extent, then, this particular trend was intuitive. Intuition is a useful tool in forecasting, but forecasts should be tested against the data that exists. My forecast in this area could be considered a kind of expert intuition, to the extent that the years I have spent in augmented cognition unknowingly (or, at most, subconsciously) informed it.

What are AugCog's prospects for adopting neurotechnologies other than the three shown above? fMRI is not likely to be adopted anytime soon, nor are tDCS or TMS. There are two reasons for this that work hand-in-hand. AugCog has a "do it yourself", or DIY spirit; further, these neurotechnologies (fMRI, especially) are more costly and less wieldy to use. One exception to AugCog's typical non-reliance on fMRI is Borders, Dennis, Noesen, and Harel's (2020) study using fMRI to predict the efficacy of training in visual scene analysis.

Augmented cognition can continue to draw from various areas. Computational neuroscience has a lot to offer in its interest in neural nets. The Massachusetts Institute of Technology (MIT), through its McGovern Institute, employs the computational model of the mind in its study of neural computation. According to McGovern's website, the institute's researchers study neural nets as part of its computational neuroscience research (Computational Neuroscience - MIT McGovern Institute).

What are augmented cognition's prospects of achieving an objective ontology? Given cognition's at least partially phenomenological nature, this question makes more sense to consider epistemically (that is, knowledge-wise).

Cognition can be partially understood as rule-based when it comes to formal syntax and its mechanisms. Augmentation as a technological goal is objective since it uses the concrete tools of analysis of brain-imaging to map an already objective ontology (the brain). Cognition as computation can certainly be treated as objective, both ontologically and epistemically. But some parts of cognition, namely the mentioned experiential part, may not admit so easily (if at all) to objectification.

Much of how one should evaluate cognitive work depends on the paradigm it comes from. In AugCog, cognition is overwhelmingly studied with a computational-neurophysiological lens. In cognitive psychology, eye-tracking is a common method used to operationalize cognitive involvement in a task (like reading). In such studies, there is at least implicit theoretical leap made from a research participant's thought process to what their eyes show them. This assumption works if and only if such a subject's observable physiological state mirrors their cognitive one[4]: i.e., eyes moving correspond with thought about the stimulus they are paired with. With the proper controls in place for an experiment, it can be reasonably inferred that such mind–body coupling is taking place, and that physiological data taken is done validly. Ideally, a cognitive-behavioral approach is adopted for such studies. But if cognition is a black box that cannot be seen through, *and* a purely cognitive perspective is applied, there is reason to doubt the validity of such studies on epistemological and ontological grounds.

Neurotech AC should attempt to carve a middle path in how it augments cognition. Heidegger distinguished between two kinds of thought: meditative and calculative. Undoubtedly, the bulk of AC is devoted to bolstering the latter. Yet efforts exist via apps like Calm that aim to slow a user's pace of life down. If technology is like a bicycle for the mind (affording it more powerful computation), could it also be like a cushion for the mind? Such a middle approach could look to cognitive psychology for inspiration. Systems 1 and 2—referring to intuition and reason—have been studied to the point where the lesser-known System 3, a conceptual average of Systems 1 and 2, was more recently proposed (Slingerland, 2014).[5]

How, other than the Platonic-Freudian model I offered in Chapter 1, can we view our minds? Are they competitive landscapes for our attention? If so, AugCog may have its hands full. However one views the mind and cognition, it is important to get clear about it before attempts to engineer it are made. Similarly, more invasive brain technologies ought to be mindfully applied with

---

[4] This consideration is somewhat related to the methodology of self-report, which admits to skepticism within psychology (a field where self-report is the go-to method asked of participants completing surveys or reporting on their experience). AugCog studies in HCI can resemble social science studies structurally, so this makes sense.

[5] This may be more aptly, albeit less catchily, named System 1.5.

a good understanding of the brain and how it works. This may seem trivial, but ethically, it should not be treated as such—and morally, it must not be.[6]

Kaku (2014) discussed the interesting possibility of neuro-telepathy. It will be remembered that Kaku noted the boon of EEG being that it is quickly implementable: that is, it can quickly recognize fast-changing signaling. As he wrote, EEG science has enabled rough estimates of human thought. This has been achieved via participants placing EEG helmets with sensors on top of their heads, and having them focus on images (e.g., cars). Neural activity was then measured per photograph shown, and a "rudimentary dictionary of thought" was made that matched individuals' respective cognitions with corresponding brain scans (p. 64). Computers were trained to distinguish EEG patterns based on images shown. This is perhaps among the most interesting cases of eye-tracking EEG at work, as it helps us better understand the interaction between person and stimulus (from a behaviorist perspective, this could be said to be "what it's all about"!). Still interestingly—albeit less "spooky"— EEG has been used on participants engaged in lucid dreaming (which allowed experimenters to figure out when they reached rapid eye movement, i.e., REM, sleep). When participants have been hypnotized, EEG showed subjects had only a small amount of external stimulation. This was supposed to show that hypnosis could afford recall of repressed memory.

Data miners of the mind should be judicious, applying an *a priori* framework predicting cognitive augmentation potential when possible. A pertinent question here for augmented cognition is: Is X data mining going to yield valuable information, which could be transformed into knowledge—and, eventually, into insight? Insight is the holy grail of research. I have discussed some applications in this book, but the reader might still wonder if there is not more to be gained. As a technologist and psycho-philosopher, I believe there is. As we humans coevolve with the technology we create, we learn to live more smartly. The exact nature of the insights we are reaching toward might seem vague in the process of applying research findings. But over time, as more lives are touched for the better by technologies like those discussed in this book, the real value of insight yielded is realized.

Perhaps the most interesting line of research with respect to NNs is in the study of sentience, embodiment, and game simulation. Kagan et al. (2022) published in the journal *Neuron* a study identifying synthetic bio-intelligence. Their study is important for its investigation of neuroscience's basic unit of analysis—the neuron—in relation to the more cognitive or behavioral phenomenon of intelligence. Intelligence is usually framed as a more recent byproduct of evolution. In the context of neurocognition, cognition is usually presupposed to have emerged from an underlying neural substrate (the central

---

[6] See Sood (2021) for my own treatment of this problem in the context of HCI AC.

nervous system, CNS). An augmented cognition team I was part of conducted a game theory simulation of a more economic nature. Kagan et al.'s study is decidedly more biological, specifically biotechnological. If their perspective on single-neuron intelligence is valid (and there may be compelling reasons to believe it is!), the field of neural networking could enjoy more computer game-relevant research as artificial intelligence has!

Another interesting path forward for neurotech AC is augmenting criminal profiling. It was discussed how facial recognition is the most common application of NNs. In this arena, AC could reduce errors made by police officers who pursue suspects who are not actual threats. Logically, predictive profiling would follow a linear path from past successful profiling to current suspects with backgrounds all matching said past profilings most. This approach should help cops overcome hasty judgments about civilians by augmenting their decision-making in the moment; i.e., predictions should accompany their instincts for more balanced decisions. This opens a path for more compassionate and empathetic policing if balancing the heady AugCog approach with the officer's instinct leads to a more heartfelt choice.

Future directions of cognition-augmenting neurotechnology include educational research. If another pandemic strikes and governments decide again to send public school students home for learning, it would be helpful to know more about what is going on in this novel learning situation. It is not as if children are not already home-schooled, but the magnitude at which students took to remote learning during 2020 was unprecedented. Technologists have been aware (and likely sympathetic) for a while of the wasted opportunities in both remote learning and work. But the COVID-19 pandemic catalyzed and essentially forced these to happen. While prudent, this was not ideal, meaning that there is still much we were not prepared to learn about in augmented cognition. If students and workers gain more freedom in the future to study and work (respectively) away from their normal worksites, augmented cognition researchers[7] become more responsible for bringing to light any cognitive benefits of this. (If no such benefits are found, we may figure out how to create them!)

Someday, our cars might function as BCIs that detect how we want to drive. For instance, wouldn't it be convenient if we could simply think and will lane changes, rather than having to manually signal? This could take out some of the human error involved in driving, but given the current state of self-driving vehicles and their suboptimal risk levels, this possibility is for the mid to long term.

---

[7] There exists an overlap between augmented cognition and engineering research. Specifically, IEEE (Institute of Electrical and Electronics Engineers) research has been cited within augmented cognition (Yamakawa, 1993).

If it is true that people can change how they feel, affect can also be engineered. Thus, AugCog can extend to augmented affect, i.e., "AugAff". Can behavior also be engineered via augmentation? Indeed, since the field of AugCog already includes human performance enhancement as a goal, behavioral augmentation may be presumed. The theory of radical behaviorism includes mental activity in its view of behavior, but as a form of behaviorism, still includes less cognition than AugCog. Still, radical behaviorism could be treated as relevant to AugCog's human performance extension: indeed, it may be the most suitable framework for such that is not nominally cognitive.

If behavior B manifesting as improved performance is an example of AugCog, then a kind of B is a kind of cognition C. This presents a challenge to the view that two of psychology's fundamental constructs are entirely distinct. If they are, in fact, interchangeable to at least this extent, they might even share a basic quantitative unit (perhaps a "psy"). The relation between B and C may then be more akin to that between matter and energy in physics.

What about motivational engineering—is this solely the domain of coaching, parenting, etc.? Could brain technologies shed light on any of this? Beyond virtual assistance and (haptic) prompting, it is difficult to see how those coveted thus far can. EEG could shed light on the brain regions involved in psychological augmentation. BCI AugPsy (Sood, 2021) is also speculative at this point. Could we simulate the AugPsy brain via NNs?

A clear mind free of stray thoughts is essential to performing well in non-cognitive domains. Consider the case of active engagement with a task consisting of the disappearance of thought. Paradoxically, in this case, AugCog can be relevant in contexts that are theoretically or actually non-cognitive. Somewhat curiously, transformation of thought into behavior augments cognition not by enhancing it, but instead by turning it into something else. Increased task engagement addresses cognitive extension into performance enhancement, and efforts to boost such engagement can be called cognitively augmentative. Technology in general extends cognition, in the sense that it manifests its inventors' goals to enhance human capability: it extends a technologist's vision into empirical reality. Assume cognition and behavior are equivalent mathematically. (There is a decent reason to do so, enough to warrant an attempted proof.) If said equivalence is sufficiently similar to that between energy and mass: cognition and behavior are also equivalent in non-mathematical reality (i.e., they are equivalent in psychological reality, as mass and energy are so in physical reality[8]).

---

[8] Though augmented cognition as defined in the field is heavily physical—for instance, in its privileged inclusion of neurophysiology—it is not proper theoretically at this point to attempt to reduce cognition to a purely physical set of processes and structures. Thus, I have distinguished between physical and psychological realities in the past (see Sood et al., 2019).

Intelligence may be the first concept that comes to mind for some when they think of augmented cognition. It is easy to associate augmented cognition with enhanced intelligence. Neurotechnological AugCog could look toward studies on intelligence to see how their technologies could be leveraged to improve it. An example (albeit outside of technological augmented cognition) is Blackwell, Trzesniewski, and Dweck's (2007) study showing that belief in intelligence as changeable led to higher educational achievement for grade-schoolers.

A related area of exploration is remote therapy. Though now a more awkward time for the industry given a recent data scandal, rates of remote therapy consumption might rise if people prefer the freedom of remote versus onsite appointments. Rather than learning, augmented cognition here could be relevant to how therapists and/or clients process and experience the new therapeutic context. Again, augmented cognition comes into play to assess potential benefits of this kind of arrangement. These notwithstanding, augmented cognition practitioners could perhaps make therapy even more beneficial!

Augmented cognition could look to fictional applications of brain technologies like those discussed in this book for inspiration. For instance, the CW television series *Supergirl* features an alien character (nicknamed "Brainy") with a neural network functioning as his brain. The alien states having "recently restructured" his NN to "form many small compartments for storing sensitive information", in Season 4, Episode 10. This actually points to the importance of a concept important in cognitive science, i.e., *modularity*. Brain-scanning in cognitive neuroscience has mapped brain regions to specific functions (e.g., the frontal lobe is "in charge of" executive functioning and deliberative decision-making). Modularity refers to the fact that the mind can be segmented into distinct functions. This much is not necessary to argue against, as we commonly accept functions such as attention and decision-making as being distinct. Still, the author wonders whether modularity—not itself a unanimously accepted principle—tells the whole story of the whole brain, or whether further research should focus more on how the brain's various parts work together to build the whole *gestalt* that is everyday perception.

The gestalt theory of perception originated from the work of French philosopher Maurice Merleau-Ponty, a preeminent phenomenologist of the last century. Merleau-Ponty philosophized that perception is ordinarily a whole that people experience, not (necessarily) occurring in parts. The most impact Merleau-Ponty has had on science—rather impressively, given phenomenology's historical opposition to mainstream science—is seen in cognitive science. Specifically, the embodied paradigm is part of a broader movement known as "5E cognition". In this model, the five *E*'s of cognition denote that it is embedded, embodied, enacted, extended, and ecological. The thesis of embodied cognition is simply that mind and body are connected in such a way that cognition (as bodily beings like us know it) occurs in a specific, foundational way. It is usually at odds with modern interpretations of René Descartes' proposed dualism separating mind (or, for him, soul) from body. BCI is among the best applications demonstrating embodied cognition. *Enactive cognition* renders behavior conceptually relevant to cognition (rather than only being related to the extent that behavior is studied in psychology and augmented cognition). There is good room for innovative work to be done at the intersection of 5E cognition and neural nets. Specifically, neural net-enabled AI developers could learn from the history of AI since the 1950s (Dreyfus, 1992). Neural nets gained traction partially, given the limits of purely rule-based AI systems in their ability to operate in the real world. Fuzzy logic programming, if possible, could augment this situation. During programming, a programmer cognizes and programs their computer to carry out one or more tasks. Developing the logic of a program is cognitive prior to representing it via a flow diagram. Programming is an activity conducted for and within augmented cognition (in addition to its obvious inclusion within HCI).

From "good, old-fashioned AI" (a.k.a. GOFAI) to connectionist theory and reinforcement learning up to now, the history of AI suggests that insights from phenomenology (as forwarded by philosopher Martin Heidegger, as well as by Merleau-Ponty) could complement the project of artificial general intelligence (AGI). Neural net-enabled AI (especially if AGI or "strong AI" becomes feasible) fusing 5E cognition (as this area's methods become more empirical) with robotics presents an interesting opportunity for multidisciplinary research.[9]

The curious fact about AI is that, though a subset of cognitive science, much of its historical impetus has been behavioral. This much is consistent enough with the human performance extension goal of AugCog. The type of intelligence AI research and development has invested in has been behavioral to the extent that the intelligence in question is for a system to learn and carry out a task. In the case of NNs and deep learning, this could still leave the hidden layer as a black box with no cognition observable from the outside. But in the case of cognitive tasks like decision-making, involving the weighing of and choosing between multiple options, a cognitive process is at least implicit. Intelligence itself has cognitive aspects to it, especially in IQ testing and the logical–mathematical intelligence forwarded by Howard Gardner (see his multiple intelligences framework). Such contexts very clearly demand thinking

---

9 In a divergent context (Sood, 2022), I have philosophized on the possibility of what I termed "fully-human AI" (FHAI). This work did not involve neural nets, but did examine whether (and if so, how much) higher-level phenomena like affect, selfhood, consciousness, and sociality could be exhibited by AI. Though I deemed FHAI generally unfeasible at the time (March of 2022), with what I have theorized here along with artificial psychological intelligence, large-language models, and biotechnological engineering, AGI becomes a more distinct potentiality.

prior to knowing how to solve a problem, for instance, completing a patterned sequence. If neural nets mimic biological brains, and brains play a significant role in cognitive processes, neural nets (especially as a function of AI) should be assumed to be related to aspects of cognition. But this also leads to confronting whether it is the brain, mind, or both that think(s). Perhaps what could be best said at this point is that people think with their minds, and it is widely theorized (almost to the point of being taken for granted) that the brain is what enables thinking.[10]

Deep learning is a subset of ML (that uses deep neural networks), which is a subset of AI. Corella and Lewison (2019) noted that ML has used ANNs since the 1950s and that such ML's "performance has improved spectacularly...for applications including face and speech recognition" in recent years (p. 42). These improvements were founded on multilayer DNNs consisting of, e.g., eight neural layers in the case of Facebook's DeepFace and 22 such layers in FaceNet. Such DNNs also use "millions of parameters", e.g., "120 million in DeepFace" and "140 million in FaceNet". These layers and parameters were adjusted via "training on millions of inputs". Includable in deep learning is generative AI, part of which is LLM. Augmentation is explicitly relevant to LLM in at least the instance of Retrieval Augmentation Generation (RAG). RAG "augments" the abilities of LLMs such as ChatGPT with an "information retrieval system" (HeidiSteen, November 20, 2023).

In RAG, data sources including files and databases are used by the (Azure) AI search. RAG illustrates the relation between user experience (UX) and ChatGPT, which is mediated by an app server or orchestrator. This communicates back-and-forth with the aforementioned AI search, using a query to arrive at knowledge. The app server/orchestrator also shares a reciprocal relationship with (Azure OpenAI) GPT or ChatGPT which also involves knowledge. However, this latter relationship combines the user's prompt (rather than query) with said knowledge to arrive at GPT's response. Via its inclusion of the neural net-powered ChatGPT and app UX, RAG is at least relevant to human–computer interaction broadly: its enablement of knowledge to the user via two bidirectional pathways covers its relevance for augmented cognition. RAG does not include UI—i.e., user interface—explicitly. However, UI could be thought of as the springboard for augmented cognition via app UX.

LLM includes GPT; GPT includes GPT-4; GPT-4 includes ChatGPT. This may be the best way to clarify ML and AI's relation to one another, and it shows the depth of deep learning. It has been necessary to discuss AI in this book given its relation to NNs, but AI is simply also ubiquitous and near-impossible to ignore at the time of writing. Its current relevance is comparable to the high

---

[10] A possible thought experiment here would be to imagine a being that could think without a brain. Is this even possible? How could anyone know—never mind the being in question?

popularity of blockchain and cryptocurrency during the second half of the last decade. AI is ubiquitous, even if only in the digital world: it is (featured) in video games (officially), on social media (unofficially), on the web, in the news, on smartphones, and used as a tool in various software applications (Siri, Bing, Google Assistant, Paint).

AugCog is responsible for humans working with machines like neuro-technologies to open the door to the best future possible for all. In this way, the field can (perhaps, should) mitigate fears surrounding AI advancement. This may only apply to AugCog NN researchers, but it is substantial enough to warrant mention.

A holographic rendering of the brain with manipulable components could increase neuroscientific collaboration exponentially, if not at least multiplica-tively. This could already be implemented via virtual reality (VR), *sans* the direct collaborative aspect. The brain could be programmed into VR like any other object, albeit with more complexity to render it in its full detail. But the level of complexity chosen for such an application just needs to match the specific problem VR-rendering the brain is intended to help solve. Imaginings of such neural nets appear to great visual and cinematic effect in two recent Marvel series, viz., *The Avengers* (films) and *Agents of S.H.I.E.L.D.* (network television series).

Culturally, neuroscience has made its mark. The Japanese game develop-ment company Konami created the Neuron mobile app; this is not a brain tech, but it shows how ubiquitous brain vocabulary has become.[11] Relatedly, the Japanese-gone-global franchise Pokémon created a monster with the Ability "Neuroforce", though the monster in question (called "Ultra Necrozma") apparently had nothing to do with the brain. However, using the broad notion of AugCog used variously throughout this book, Neuron augments cognition. It does so by allowing players to calculate scores ("Life Points"), roll a die, flip a coin, listen to music (which can augment the "Dueling" experience), and enjoy a satisfying user interface (UI). All of this leads to the augmentation of user experience (UX). Users select experience and interfaces—both UI and UX are important practical staples of the broader HCI landscape.[12]

The idea of humanoid artificial intelligence is not new, but the implication of a neural net powering its mind goes a step further than popular discourse. The project OpenWorm demonstrated artificial intelligence in a robot powered by a neural network simulation of a worm's brain (Seeker, 2018, January 11, 1:55-2:00). Worm brains are orders of magnitude simpler than human ones, but

---

[11] One possible reason for the app's name being such, other than to capitalize on what I call "neu-rohype", is that the app serves as a centralized technology solution for players of the *Yu-Gi-Oh!* Trading Card Game (TCG).

[12] Can neurotechnologies augment such selections? What about the experiences themselves?

increasingly, the divide between reality (as demonstrated through science and engineering) and science fiction is blurring.

Can neurotechnology in AugCog allow its practitioners and beneficiaries (including experiment participants) to experience more fully the important phenomena of joy, happiness, freedom, love, and wisdom? Related to such topics is the sense of community neurotechnology can foster (which was already discussed in Chapter 3). NeuroTechX noted that: "To become a practical reality, brain technology needs to be supported by a diverse and dynamic community of experts, entrepreneurs, and enthusiasts" (NeuroTechX, 2023).

Beyond the scope of this book is how neurotechnological AugCog could help shore up other areas. Closely related conceptually (if not necessarily philosophically) to HCI is the newer field of *postphenomenology*. Inspired by Martin Heidegger's interest in and philosophy of technology, postphenomenology is defined as the study of human-technology relations (Rosenberg & Verbeek, 2015). Technically, HCI would be a subset of this, even if there is good reason to define postphenomenology otherwise (e.g., more broadly as a human science of what is beyond experience—or a philosophy of what could be so). More simply, could neurotechnological AugCog help develop the phenomenology of thinking? One way it could is to exploit its tie with UX. Since UX is about a kind of experience (a consumer's or computer user's), its study could be classified as a kind of phenomenology. Such a phenomenology would fit well under postphenomenology as just presented, helping unite it with HCI in a more disciplinary way. This could expand neurotech AC's purview into postphenomenology (which could also bring it closer to embodied and enactive cognition, an area that has drawn explicitly from Merleau-Ponty's phenomenological philosophy).

The concept of a brain disorder is one that could benefit from a skeptical stance. Since all knowledge of the brain based on neurotechnology is fairly recent, the sense of "disorder" has to be clear. One dichotomous (and common) way to analyze the brain is in terms of its structure and function. Brain scans can show irregularities in brain activity (and, by extension, function). When these deviate from normal brain activity, a disorder may be at play, structurally and functionally. But irregularity alone could mean extraordinary functioning, an ideal of AugCog (if brain function is coupled with cognition). Brain disorders need to have a traceable cause, such as a TBI, in order to be considered such.

Another way neurotech AC might help is to make computers more sensitive to context. For instance, as I wrote the above paragraph (on a Google Doc), a suggestion was made to alter a phrase ("is to" to "be to") that ignored the broader context of the sentence. Making this alteration would not have made sense grammatically, but could neurotech AC help here? This seems a case showing such AC's limits. This might be a project for AC in general, but it does not seem like neurotech would have any bearing on this particular problem. It then becomes worth asking what problems neurotech AC *is* suited for.

An interesting adaptation of ChatGPT is ChatCBT. CBT—i.e., cognitive-behavioral therapy—is conducted so that a patient learns to recognize their limiting beliefs and convert them into effective action. CBT done properly augments cognition (albeit non-technologically, except for the case of remote therapy) to the extent that human performance is boosted by way of initial engineering of one's thought. An NN-based application like ChatCBT is also appropriate as it reflects the early history of AI and the psychotherapy chatbot ELIZA. Turning irrational cognition into effective behavior is certainly a good example of augmented cognition, and being able to do so using an app designed for such represents a good case study for future AugCog focus.

One could imagine AugCog leading to solving difficult theoretical problems and perhaps global grand challenges. Ideally, on the macro-level, AugCog is fulfilled by enabling achievement of the United Nations' Sustainable Development Goals (SDGs). Where it is implausible to do so given the desired deadline (e.g., 2030 or 2050), AugCog could contribute a more realistic timeframe or make SDG achievement adequately tractable. If AugCog can use neurotechnologies like those discussed in this book to solve the world's most pressing problems and lift cognition (and behavior) beyond its current level, it will have fulfilled its duty to the world, and the techs used will have proven their ultimate utility. AugCog is already a global entity, meaning it has the potential to solve global problems—all that would be left (at least, as far as this book is concerned) is to figure out the details of where and how neurotechnologies can be used for such purposes.

One limitation of my trend analysis is that I did not discuss AugCog neurotechnologies prior to 2023. They were included in my term frequency analysis, but I did not deem them worth going into depth about because they are not as current as last year. My descriptions of 2023 neurotech AC studies were done to give the reader a good idea of the most relevant work in the area. Some of the empirical work and applications I've discussed have inevitably left some vagueness in their descriptions. This is primarily because I did not have direct contact with them, but also because—even if I had—you may not have, either. Treat descriptions of the studies and technologies in this book as an impetus to go on your own journey, discovering how neurotechnology can improve thought!

We change the future by changing the present. Future trends, then, are determined by current ones. We have free will to shape the present, but it needs to be understood that doing so will impact the future to come, as determinism is also true. So, AC neurotech development and application should be done mindfully. The future of AugCog looks bright, and the three neurotechnologies discussed in this book—electroencephalography (EEG), brain–computer interface (BCI), and neural networks (NN)—should help make that the surest trend of any identified here.

REFERENCES

Blackwell, L. S., Trzesniewski, K. H., Dweck, C. S., Implicit theories of intelligence predict achievement across an adolescent transition: a longitudinal study and an intervention, *Child Development*, 78(1), 246–263, 2007.

Borders, J. D., Dennis, B., Noesen, B., Harel, A. Using fMRI to Predict Training Effectiveness in Visual Scene Analysis. In: Schmorrow, D. D., Fidopiastis, C. (eds) *Augmented Cognition. Human Cognition and Behavior. Lecture Notes in Computer Science*, vol. 12197, Springer, Cham, 2020. https://doi.org/10.1007/978-3-030-50439-7_2

Computational Neuroscience – MIT McGovern Institute. *mit.edu.* https://mcgovern.mit.edu/research-areas/computational-neuroscience/

Corella, F., Lewison, K., Biometrics. In: Moallem, A. (ed), *Human–Computer Interaction and Cybersecurity Handbook,* Boca Raton, CRC Press, 2018.

Dreyfus, H., *What Computers Still Can't Do: A Critique of Artificial Intelligence*, The MIT Press, Cambridge, MA, 1992.

HeidiSteen, Retrieval Augmented Generation (RAG) in Azure AI Search, *Microsoft,* RAG and generative AI - Azure AI Search | Microsoft Learn, 2023, 20 November.

Kagan, B. J., Kitchen, A. C., Tran, N. T., Habibollahi, F., Khajehnejad, M., Parker, B. J., Bhat, A., Rollo, B., Razi, A., Friston, K. J. In vitro neurons learn and exhibit sentience when embodied in a simulated game-world. *Neuron*, 110(23), 3952–3969, 2022.

Kaku, M., *The Future of the Mind*, Doubleday, New York, 2014.

NeuroTechX, NeuroTechX – Slack Group for Neuroscience, Hive Index, 2023. https://thehiveindex.com/communities/neurotechx/

Rosenberg, R., Verbeek, P.-P. *Postphenomenological Investigations: Essays on Human-Technology Relations.* Lexington Books, Lanham, MD, 2015.

Seeker. Scientists Put the Brain of a Worm Into a Robot… and It MOVED, *YouTube,* 2018, 11 January. https://www.youtube.com/watch?v=eYS7UIUM_SQ

Slingerland, E., *Trying Not to Try: The Art and Science of Spontaneity*, Crown, New York, 2014.

Sood, S. Holarchic HCI and Augmented Psychology ("AugPsy"). In: Schmorrow, D. D., Fidopiastis, C. M. (eds) *Augmented Cognition. Lecture Notes in Computer Science (LNAI)*, vol. 12776, Springer, Cham, 2021. https://doi.org/10.1007/978-3-030-78114-9_22

Sood, S. Could IBM's Deep Blue Chess Program Feel Triumphant?. In: Arai, K. (ed) *Advances in Information and Communication. Lecture Notes in Networks and Systems*, vol. 438. Springer, Cham, 2022. https://doi.org/10.1007/978-3-030-98012-2_55

Sood, S., et al. Holarchic Psychoinformatics: A Mathematical Ontology for General and Psychological Realities. In: Schmorrow, D. D., Fidopiastis, C. (eds) *Augmented Cognition. Lecture Notes in Computer Science*, vol. 11580, Springer, Cham, 2019. https://doi.org/10.1007/978-3-030-22419-6_24

Storey, V. C., Lukyanenko, R., Castellanos, A., Conceptual modeling: topics, themes, and technology trends, *ACM Computing Surveys,* 55(14), 1–38, 2023.

Yamakawa, T., A fuzzy inference engine in nonlinear analog mode and its application to a fuzzy logic control. *IEEE Transactions on Neural Networks*, 4(3), 496–522, 1993.

Glossary

- **Augment**—to enhance, improve, and make better, especially using technology.
- **Augmented cognition (*groups*)**—researchers and industry professionals interested in enhancing cognitive capabilities using adaptive technologies.
- **Augmented cognition (*phenomenon*)**—enhanced thought via technological integration with the subject that can be extended into behavior and performance.
- **Brain**—a complex web of neurons interacting with one another via neurotransmitter signals.
- **Brain–computer interface**—a technological connection made between a person and a computer, with the aim of recording their neural activity to be translated into inference about their cognition (e.g., intention).
- **Electroencephalography**—a method of recording high-level brain signals, through use of a helmet and electrodes, to be transformed into data analyzable via devoted software.
- **Enactive cognition**—part of the 5E cognitive framework, refers to thought as a function (at least partially) of the environment or behavior.
- **Forecasting**—a practice between speculation and prediction to better understand future scenarios.
- **Human–computer interaction**—a discipline at least partially containing augmented cognition and including psychology, linguistics, philosophy, artificial intelligence, neuroscience, and cognitive science.
- **Neural network**—a biological or artificial nervous system, with the latter being used as a simplification of the former to enable computational functions (such as facial recognition).
- **Neurotechnology**—any hardware or software that models, scans, and/or affects the brain or (part of) the nervous system
- **Sensory substitution**—David Eagleman's idea that a deficit in one sense (such as sight) can be compensated for by another, more functional sense (e.g., touch)

- **Technology**—includes computers and the ensemble of hardware and software that make up what is discussed in this book
- **Trend**—consistency of use or participation over time in a given domain.

Index